T0134560

Lecture Notes in Networks and Systems

Volume 162

The series "Lecture Notes in Networks and Systems" publishes the latest developments in Networks and Systems—quickly, informally and with high quality. Original research reported in proceedings and post-proceedings represents the core of LNNS.

Volumes published in LNNS embrace all aspects and subfields of, as well as new challenges in, Networks and Systems.

The series contains proceedings and edited volumes in systems and networks, spanning the areas of Cyber-Physical Systems, Autonomous Systems, Sensor Networks, Control Systems, Energy Systems, Automotive Systems, Biological Systems, Vehicular Networking and Connected Vehicles, Aerospace Systems, Automation, Manufacturing, Smart Grids, Nonlinear Systems, Power Systems, Robotics, Social Systems, Economic Systems and other. Of particular value to both the contributors and the readership are the short publication timeframe and the world-wide distribution and exposure which enable both a wide and rapid dissemination of research output.

The series covers the theory, applications, and perspectives on the state of the art and future developments relevant to systems and networks, decision making, control, complex processes and related areas, as embedded in the fields of interdisciplinary and applied sciences, engineering, computer science, physics, economics, social, and life sciences, as well as the paradigms and methodologies behind them.

**** Indexing: The books of this series are submitted to ISI Proceedings, SCOPUS, Google Scholar and Springerlink ****

More information about this series at http://www.springer.com/series/15179

Kamel Boussafi · Jean-Pierre Mathieu · Mustapha Hatti
Editors

Social Innovation and Social Technology

Enterprise-New Technology Synergy

Editors
Kamel Boussafi
École Supérieure de Commerce
Tipasa, Algeria

Jean-Pierre Mathieu
CEPN-CNRS
Paris 13 University
Villetaneuse, France

Mustapha Hatti
Equipements Solaires, EPST-CDER
Unité de Développement des
Bou-Ismail, Algeria

ISSN 2367-3370 ISSN 2367-3389 (electronic)
Lecture Notes in Networks and Systems
ISBN 978-3-030-60932-0 ISBN 978-3-030-60933-7 (eBook)
https://doi.org/10.1007/978-3-030-60933-7

This Springer imprint is published by the registered company Springer Nature Switzerland AG
The registered company address is: Gewerbestrasse 11, 6330 Cham, Switzerland

Contents

About the Editors

Prof. Dr. Kamel Boussafi is the Director of the Ecole Supérieure de Commerce at the Pole-University of Koléa, Tipasa in Algeria, director of several research projects including CNEPRU; PHC TASSILI; REDSIEM Lab, Consultant trainer at INPED in the field of business intelligence and innovation management. Responsible for finance training at ESC-Koléa, Tipasa Algeria and LMD training at the Algiers Tourism School. Supervised several doctoral theses and several master's theses. Chaired and reviewed several doctoral theses. Author of several publications, organizer of several international onferences and Symposia.

Prof. Jean Pierre Mathieu is a Professor and Senior Researcher in Marketing and Management Sciences (PhD of University Pierre Mendes France, Grenoble and "Habiliation à diriger les recherches" from the University of Paris XIII). Between 1992 and 1994, I was an Assistant Professor in University of Savoie (France). In 1994, I joined the Audencia Group as the Director of the professional development activities (also known as executive education program). I held this position until 2004. Since 2004, I have developed and supervised several academics and scientific programs. Among the different actions I led, I created the first Master of Marketing and Design. This new program has been the first in Europe to combine lectures and practical sessions dealing with management, engineering and design skills. From a research point of view, I supervised several PhDs focusing on similar challenges, aka the marketing of the design and the innovation. In addition to multiple publications in conferences and journals, the different results we obtained, have led to the creation of a new paradigm: Marketing of the design. Leveraging more than 20 years of experiences from the Audencia group and several worldwide collaborations (USA, Asia, Africa), my actives are now focusing on the consulting of large companies and academics institutions.

Dr. Mustapha Hatti was born in El-Asnam (Chlef), Algeria. He studied at El Khaldounia school, then at El Wancharissi high school, obtained his electronics engineering diplomat from USTHB Algiers, and his post-graduation studies at USTO –Oran (Master's degree and doctorate). Worked as research engineer, at CDSE, Ain oussera, Djelfa, CRD, Sonatrach, Hassi messaoud, CRNB, Birine, Djelfa, and senior research scientist at UDES / EPST-CDER, Bou Ismail, Tipasa, "Habilité à diriger des recherches" HDR from Saad Dahlab University of Blida, Algeria. Actually, he is Research Director in renewable energy. Since 2013, he is an IEEE senior member, he is the author of several scientific papers, and chapter books, and his areas of interest are smart sustainable energy systems, innovative system, electrical vehicle, fuel cell, photovoltaic, optimization, intelligent embedded systems. President of the Tipasa Smart City association. An eBook editor at International Springer Publishers, guest editor and member of the editorial board of the journal Computers & Electrical Engineering, he has supervised, examined and reviewed several doctoral theses and supervised master's degrees. He organizes the conference entitled Artificial Intelligence in Renewable Energetic Systems.

The Associative Boom in Algeria: Reality or Democratic Illusion?

Sonia Bendimerad[1](✉), Amina Chibani[2], and Kamel Boussafi[2]

[1] Spaces and Societies Research Unit (Espaces et Sociétés ESO 6590 CNRS), University of Angers, École supérieure de commerce d'Alger, PHC Tassili Research Team, Angers, France
sonia.bendimerad@univ-angers.fr

[2] Management, Governance Innovation, and Organizational Performance Research Unit, École supérieure de commerce d'Alger, PHC Tassili Research Team, Angers, France

Abstract. This article studies the development of the associative sector in Algeria. After providing an introduction to the existing legal framework and the recent changes in the legislation, the authors examine the initiatives implemented by the government regarding associations, as well as the economic, social, and societal problems that can hamper them. Based on a sample of 145 national associations, the authors have conducted an exploratory study to analyze the impact of government actions on the distribution, growth, and diversity of the Algerian associative sector. Their study shows that the policy of democratization proposed by the Algerian government to foster the associative sector is a veil of illusion masking mechanisms of monitoring and of restriction of the freedom of the associative sector.

Associations in Algeria (known as *Jam iyyat*) are private, generally non-profit organizations, independent from the state, whose creation and operations are nevertheless regulated by legislation at the national level. Between the mid-1980s and 2006, Algeria experienced the failure of its political liberalization process and underwent a prolonged economic collapse, both of which contributed to a dramatic decrease in the population's quality of life. During this same period, which was marked by political unrest and rising violence, associations grew in number, to the point that Algeria now has one of the highest concentrations of associations of any country in the Middle East and North Africa region (Liverani 2008). There is, however, a distinct lack of information or analysis on the growth of the Algerian associative sector (Charif and Benmansour 2011), making it one of the most inscrutable aspects of the country's recent political history. Our study aims to fill this gap. The weakness of the existing literature on the associative boom in Algeria creates two problems. First, studies of this sector generally do not explore the concrete reality of those associations that are truly active within it. Second, existing studies tend to assume that the existence of an associative fabric is a prerequisite for democracy, without actually going into what kinds of changes fulfilling this condition would bring about.

By using Algeria as a case study, this article aims to provide a better understanding of the role played by the country's political economy in the development of its associative movement. In this study, terms such as "associative life," "associative fabric," and "associative sphere" are used to refer to all the associations that are active within the country. The term "associative movement" refers to the

K. Boussafi et al. (Eds.): MSENTS 2019, LNNS 162, pp. 1–15, 2021.
https://doi.org/10.1007/978-3-030-60933-7_1

institutional space that they occupy, as well as the social and material practices that characterize their activities (Law 90-31 1990).

Based on the existing literature, this study asks the following questions: What recent changes have there been within Algeria's associative movement? What impact have the reforms proposed by the state had on the way associations operate? Have these reforms supported the contributions that associative life makes to democracy? If so, is the "associative boom" a reality or an illusion?

In the first section, we will review the natural, cultural, religious, historical, and economic factors that have more or less directly shaped the emergence of the associative movement in Algeria. Then, we will use several examples to analyze the various reforms implemented by the state to improve the lives of disadvantaged populations. Finally, based on the chronological description of the successive changes to the legal framework, we will analyze the role of the Algerian state in the creation and operation of associations.

1 The Structure and Peculiarities of the Algerian Associative Fabric

The reality of associations in a country cannot be understood independently of that country's geographical, cultural, historical, and economic context (Ben Néfissa 2002). Algeria's geographical context, namely its location in the heart of the Maghreb, makes the country at once African, Middle Eastern, and Mediterranean. The largest country in Africa in terms of area (2,381,741 km^2), Algeria has a young population and abundant natural resources (iron, oil, and natural gas). Despite these riches, some entire areas in the Sahara and in the mountainous Kabylie region remain isolated. This has driven local populations, which are left to fend for themselves, to develop various practices of mutual aid.

1.1 The Origins of the Associative Movement: Historical and Cultural Factors

Algerian families and tribal groups have long practiced their own specific forms of solidarity, with various actions that require participation and contribution from everyone. Religious organizations (*Zawiya*) aimed to strengthen social bonds by fighting against certain forms of exclusion and vulnerability. Islam, which is the religion of 99% of the population (Cherbi 2017), is perceived in Algeria not only as an expression of faith, but also as a state of mind that calls for the peaceful coexistence of different social groups, with the goal of promoting solidarity. Several forms of religiously-inspired solidarity, such as *wakfs*[1] and *tiwizas*[2] (Bozzo and Luizard 2011) were institutionalized by the Algerian Ministry of Religious Affairs and *Wakfs* via two solidarity funds (the *Zakat*

[1] A kind of donation in perpetuity from an individual for religious, charitable, or public utility purposes.

[2] In a family, village, or tribal context, *touiza* or *tiwizi* refers to a form of community cooperation or development based on reciprocal giving. Because it is based on reciprocity, this practice of solidarity is the most widespread in Algeria. It helps to mobilize available human resources and to pool material means for projects that will help families in need.

al-Fitr[3] and the *zakat*). Islam also holds that every human life is sacred and that since material goods come from God, people should give some of what they receive to the poor, hence the obligatory nature of almsgiving or *zakat* (Ducellier and Micheau 2016). The associative movement, which has only continued to grow, is therefore anchored in the religious traditions of the Algerian people (Liverani 2008; Merad Boudia 1981).

The first Algerian associations were created at the beginning of the twentieth century, after the promulgation of the French Law of Associations of 1901. In Algeria, this law served as the framework for the development of a rich and diverse associative fabric (Dris-Aït Hamadouche 2017), made up of three kinds of organizations: mixed associations, which included both Algerians and Europeans and which gravitated around the labor movement; associations made up of European colonists (cooperatives, social clubs, and sports clubs); and associations with an exclusively Algerian membership, such as Muslim charity associations and sociocultural and educational associations. These structures played an important role in mobilizing young Algerians against French colonialism. The Law of Associations of 1901 remained in effect after Algeria gained its independence in 1962 and was only repealed by decree in 1971.

After independence, a planned, centralized, state-run socialist economy was put in place, in which the public sector, with its large monopolies, was omnipresent (Adel and Guendouz 2015). The way in which this economic model focused more on social benefit than economic performance contributed to the collapse of oil prices in 1986, which led to an unprecedented resource crisis (Talahite and Hammadache 2010). This crisis revealed the weakness of the Algerian economy, with dramatic social consequences. Poverty and unemployment rose sharply, while purchasing power plummeted (Ould Aoudia 2006). Riots broke out in October 1988, leading the country to enact a series of economic reforms. The process of opening up the economy and the state's need to limit material and human support due to the ongoing economic crisis both contributed to the revitalization of the associative movement. This led to the passage of a freedom of association law, which was ratified on December 4, 1990.

1.2 Public Policy and Social Cohesion

In an attempt to resolve the economic crisis, the Algerian state took measures to ensure social cohesion, which have barely been modified since their creation. According to a 2016 report by the United Nations Development Programme (UNDP), the policies that were adopted to assist disadvantaged populations can be summarized as follows:

– Created in 1992, the Solidarity Allowance (Allocation forfaitaire de solidarité; AFS) and the Compensation for Activities in the General Interest (Indemnité pour une activité d'intérêt général; IAIG) are the foundation of the "social safety net." Outside of these two schemes, social assistance programs also include material and financial actions, social protections, and the care of people with disabilities in specialized facilities. Assistance is also provided to families who take in orphans (study carried out at the Ministry of National Solidarity, 2016).

[3] *Zakat al-Fitr* is charity given to the poor at the end of Ramadan.

- The creation in 1994 of institutions to tackle unemployment, including the National Unemployment Insurance Fund (Caisse nationale d'assurance chômage; CNAM) and the National Employment Agency (Agence nationale pour l'emploi; ANE), which aimed to help young people find long-term careers. Even though unemployment currently stands at around 13.2% (study carried out at the Office national des statistiques, 2018), both Algerian employment authorities and the country's young people themselves continue to underestimate the potential of the associative sector. However, working in an association helps one to develop various skills and areas of expertise (administrative and financial management, project coordination, communication, and so on). There are also many different areas of activity that correspond to government-targeted sectors that are full of potential for increasing growth and creating jobs (Zoreli 2016). These areas include renewable energy, professional training, digital technology, tourism, sports and leisure, and even health care. There are already many Algerian associations that focus primarily on these areas.
- In 1996, the Social Development Agency (Agence de développement social; ADS) was created. This agency provides assistance to those in extreme poverty and aims to improve public solidarity. In 1997, the Ministry of National Solidarity was charged with managing solidarity and social action programs to combat poverty and social exclusion.
- The state's management of social issues in its development programs was formalized with an Economic Recovery Support Program (Plan de soutien à la relance économique; PSRE), a Proximity Program for Rural Development (Programme de proximité de développement rural; PPDR), and a Proximity Program for Integrated Rural Development (Programme de proximité de développement rural intégré; PPDRI), among others.[4]

However, the effectiveness of these long-term measures remains dependent, on the one hand, on all social partners and the associative movement in the implementation of public employment promotion schemes, and, on the other hand, on compliance with conditions of rigor, fairness, and transparency when assistance is granted and provided to different categories of beneficiaries (Akesbi 2017).

2 The Legal Framework of the Algerian Associative Movement

The period that followed the 1990s was marked by a weakening of democratic institutions—a situation that has continued to this day.

2.1 The 1990s: A "Restricted" Opening up of Associative Activities

Despite the existence and application of a more liberal law in 1990, the new legal framework for associations remained fairly vague. As Laurence Thieux (2009) notes, "some provisions allowed the public authorities to retain mechanisms for monitoring

[4] We can better understand the logic behind intervention by examining the National Social Budget (Budget social de la nation; BSN), which is the Algerian state's preferred instrument of social action.

and limiting the exercise of freedom of association."[5] This led to the following strict criteria: associations could only be created by groups of at least fifteen adults, of Algerian nationality, all of whom had to be in full possession of their civil rights, and none of whom could have displayed behaviors contrary to the interests of the fight for national independence.

On a different note, this law also seems to show suspicion of associations' methods of obtaining financing, especially from abroad. For this reason, it was determined that donations and contributions from foreign organizations had to be approved in advance by the competent public authority, which would verify the source of the donation and its connection to the association's stated goals.

As Bachir Senouci (1999) has noted, "during the first years of the civil war, the most turbulent period in Algerian political history, the number of associations grew considerably." In the 1990s, the state created spaces for consulting with associations in order to reinforce its own domestic legitimacy: examples include the National Dialogue Conference (Conférence du dialogue national) in 1995 and the National Transition Council (Conseil national de transition), with 85 of the council's 100 seats reserved for association members. Other types of associations that were more combative and independent of the state also appeared, signaling a wider desire for equality, justice, and human rights.

At the end of the 2000s, Algerian associations became active in many different areas, in particular environmental protection, health care, consumer rights, sports, and protecting the country's historical and architectural heritage. They offered services and assistance to disenfranchised populations in neighborhoods, cities, and regions. Some even focused on specific groups, such as abandoned children or people with reduced mobility.

2.2 The 2000s: Algerian Associations and Their Freedom "Under Surveillance"

Following the Arab Spring, the Algerian regime launched a whole host of initiatives to channel the wave of protest movements that had sprung up in cities across the country in January 2011 (Dris 2013; Volpi 2014; Mokhefi 2015). President Bouteflika even gave a speech on television on April 15, 2011 (Dris-Aït Hamadouche 2012). Parliament passed several of the promised reforms as new laws, including a new regulatory framework for associations, which was published in the state's Official Journal on January 15, 2012. This law, made up of seventy-four articles (*madda*), is divided into four main sections (*Bab*).

The first section covers the creation of associations. The law of 2012 distinguishes between three types of associations: national associations; associations on a *wilaya*[6]

[5] **Translator's note**: Unless otherwise stated, all translations of cited foreign-language material in this article are our own.

[6] By definition, the *wilaya* is the division of territorial government above the local level. It is administered by a *wali*, who holds executive power, along with all ministerial departments and agencies. The intermediary level transposes the *wilaya*'s mode of operation onto smaller territories; it is based on *daïras*, or districts, which are linked to subdivisions. Each *daïra* is administered by a district chief, who oversees several communes. Finally, at the bottom of the hierarchy are the communes, which are administered democratically by elected councilors and a mayor representing the local population.

(provincial) level; and associations on a community (district) level. The number of members required to create an association varies: national associations require a minimum of twenty-five founding members from at least twelve different *wilayas*; provincial associations require a minimum of fifteen founding members from at least two different communes; and community associations need at least ten members to get off the ground. The new legislation not only retains associations' obligation to notify the public authorities of any changes to their bylaws, executive bodies, or financial situation, but it also requires that associations submit a copy of the minutes of their meetings, as well as their annual and financial reports after each ordinary or extraordinary general meeting, within thirty days of their approval.

This law defines associations as "the grouping of natural and/or legal persons on the basis of a fixed-term or permanent contract. These persons pool their knowledge and means, as volunteers and on a non-profit basis, in order to promote and encourage activities in certain areas, in particular in professional, social, scientific, religious, educational, cultural, sports, environmental, charitable, and humanitarian contexts." Any legal adult therefore has the right to found an association in Algeria, and they must follow the seven steps below (Fig. 1):

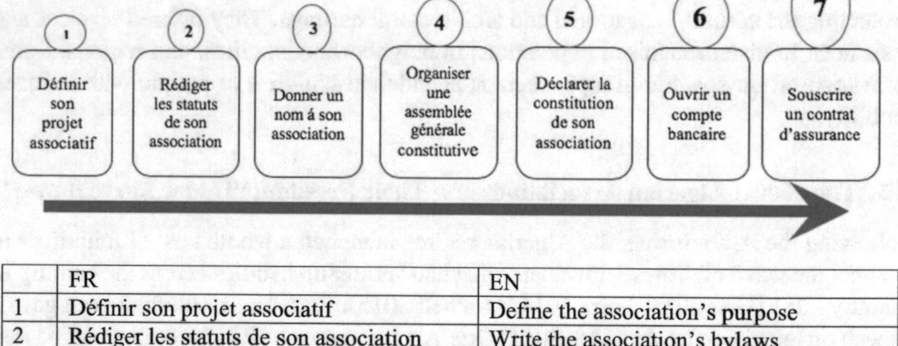

	FR	EN
1	Définir son projet associatif	Define the association's purpose
2	Rédiger les statuts de son association	Write the association's bylaws
3	Donner un nom à son association	Give the association a name
4	Organiser une assemblée générale constitutive	Organize a constituent general meeting
5	Déclarer la constitution de son organisation	Declare the creation of the organization
6	Ouvrir un compte bancaire	Open a bank account
7	Souscrire un contrat d'assurance	Take out an insurance policy

Fig. 1. The process for creating an Algerian association Source: Diagram created by the authors using data from the Ministry of the Interior and Local Authorities (2018).

The second section of the law of 2012 discusses financing methods. Financing may come from members' dues, community activities, donations, and grants from governmental bodies. However, associations are required to maintain a single account, either with a bank or a public financial institution. Also, as was the case with the previous legislation, associations are still forbidden from receiving any form of subsidy, donation, or grant from political parties and from helping to finance these political parties. The purpose of this is to separate these two kinds of entities in their objectives, definition, and operations.

The third section covers associations that are led in whole or in part by foreigners, with headquarters overseas, that are authorized to operate within Algeria. These organizations are overseen by three governmental institutions: the ministries of the interior, of foreign affairs, and of the sector in question. However, foreign associations may have their operating authorization revoked if they pursue activities other than those defined in their bylaws, if they clearly interfere with Algeria's internal affairs, or if their activities are likely to threaten Algeria's national sovereignty, the established institutional order, national unity, public order, or the cultural values of the Algerian people.

The fourth section of the law is devoted to "special associations," which may include religious associations, foundations (*El-mouassassat*), social clubs (*El Widadiyate*), and student or sports associations (*El Itihadat Toulabiya wa Riyadiya*).

After the law of 1990 was passed, it was further modified to gradually and moderately liberalize the associative sector. However, after analyzing the law of 2012, we can confirm that, while the new regulations are more elaborate and more complex than the previous ones, they also clearly contain newly introduced monitoring mechanisms. As Bachir Dahak (2014) has shown, this law restricts and penalizes the exercise of freedom of association. The Algerian Human Rights League (Ligue algérienne des droits de l'homme; LADDH) and other civil society organizations have continued to call for new regulations that are more in line with international standards.

2.3 2018: A New Associations Bill Aiming for More "Flexibility"

On January 31, 2018,[7] in response to calls from several associations, the Minister of the Interior, Local Authorities, and Territorial Planning, Noureddine Bedoui, announced that the new associations bill would include several reforms aimed at making the conditions and procedures for creating an association more flexible.[8] In particular, these reforms included reducing the required number of founding members, removing some of the administrative documents needed in the application, and making it easier to declare and seek authorization for an association's activities. The minister highlighted the right to create associations and their right to pursue their activities freely. The goal was to expand this field to include human rights and the promotion of citizenship, given their importance in society.

[7] "Projet de loi relatif aux associations: "souplesse" dans les procédures de constitution d'associations," *Algérie Presse Service*, February 1, 2019.

[8] Recent anti-government protests have slowed the process for implementing this law.

The bill thus gives associations the right to appeal decisions to refuse the creation of a new activity, as well as the ability to seek out sources of financing. It also allows for the twinning of associations that have the same objectives and that work in the same area.

3 The Realities of Algerian Associations

Data collection
In order to study the actual consequences of the state's various actions on associations' activities, we used Algeria's central statistics institution: the National Office of Statistics (Office national des statistiques; ONS). This body is charged with collecting information about all natural and legal persons using a unique and confidential statistical identification number (numéro d'identification statistique, or NIS). For associative structures, this NIS is required in order to:

- request subsidies from the state or local authorities;
- hire employees;
- pursue profit-generating activities.

Our sample of 145 structures is made up of Algerian associations that meet the three criteria for receiving an NIS. We decided to use analytical observation techniques to review this data, based on four approaches:

- describing the geographical situation of the associations in the sample;
- studying the change in the number of associations created between 1974 and 2017;
- collecting observations about Algerian associations' primary areas of activity;
- building a corpus of examples of associations that stand out by virtue of their activities, namely the associative movement by and for students, as well as that for women's rights.

3.1 Overview of the Associative Sector in Algeria

Before the political reforms of 1988, there were almost 12,000 associations officially registered with the Ministry of the Interior and Local Authorities. Ten years later, this figure had risen to 57,400, with 1,000 registered at the national level and 56,000 registered locally. This phenomenon is surprising, in the sense that the associative boom occurred during the socially devastating period of the civil war (Derras 2007).

According to official data from the Ministry of the Interior, in 2017 there were around 96,150 associations registered on the national level, including nearly 15,800 religious associations and 5,134 local associations. However, as Arab Izarouken (2014) has shown, a rising number of associations is not a reliable indicator of the dynamism of Algerian civil society. According to Izarouken, the number of associations on the official register is much greater than the number of structures that are truly active.

The table below shows the diverse array of areas of activity pursued by one thousand national associations (Table 1).

Table 1. Areas of activity in the Algerian associative sector

Area of activity	Number of associations
Professional sector (trade, professional integration, and consumer rights)	213
Healthcare sector	151
Culture and tourism sector	143
Youth and sports sector	142
Science and technology sector (for training and education)	49
Social sector (defense of the rights of women and veterans, political and religious associations)	23
Other	279

Source: Table created by the authors using data collected from the Ministry of the Interior (2018).

3.2 An Essentially Urban Phenomenon

Mapping out the associations that we identified shows that they are primarily located in or near large coastal cities. More than 80% of the 145 associations studied are active in urban areas. Northern Algeria and the *wilaya* of Algiers together account for more than 50% of all associations. These associations are rarely found on the outskirts of cities and are mostly located in city centers. Associations are also more common in regions where modes of community organization already exist (Kabylie and M'zab, for example). The associative movement is not evenly distributed across the country's territory; it is more present in the center and east of the country than the west (Fig. 2).

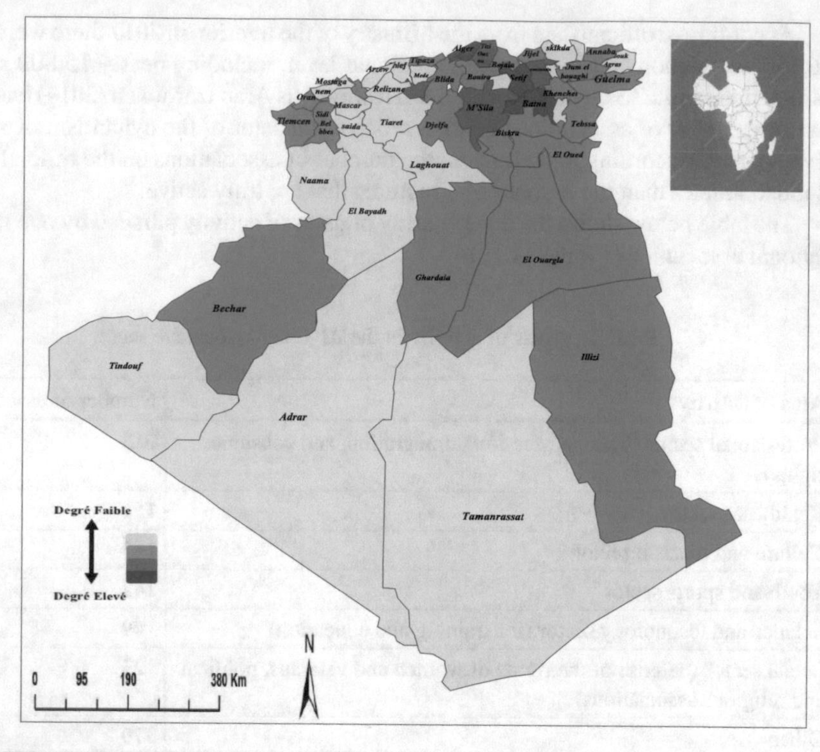

FR	EN
Densité élevée	High density
Densité faible	Low density

Fig. 2. Geographical distribution of the 145 Algerian associations in our study sample Source: Map created by the authors using sample data from the National Office of Statistics (2018).

3.3 Growth of the Associative Movement

Our study shows that the "associative boom" (Kadri 2012; Mihoubi 2014) is in fact inextricably linked to the various state reforms of this sector. As the following figure shows, there was a significant increase from 2012 onward (after the new law was passed). However, we also note decreased growth starting in 2014, which continued through 2017. This was due to the many mechanisms in this law for monitoring and dominating how Algerian associations are created and seek subsidies, which had a negative effect on the longevity of their activities (Fig. 3).

3.4 Breakdown by Sector of Activity

Our analysis confirms the fact that most Algerian associations focus their activities on cultural, social, and environmental issues—in other words, in sectors where the state has an interest in supporting public action. On the other hand, human rights associations and associations that are generally active in more politically sensitive sectors are in the clear minority (anti-corruption associations, for example). Still, as Ahcène Amarouche (2012) has highlighted, some organizations—such as feminist associations and associations

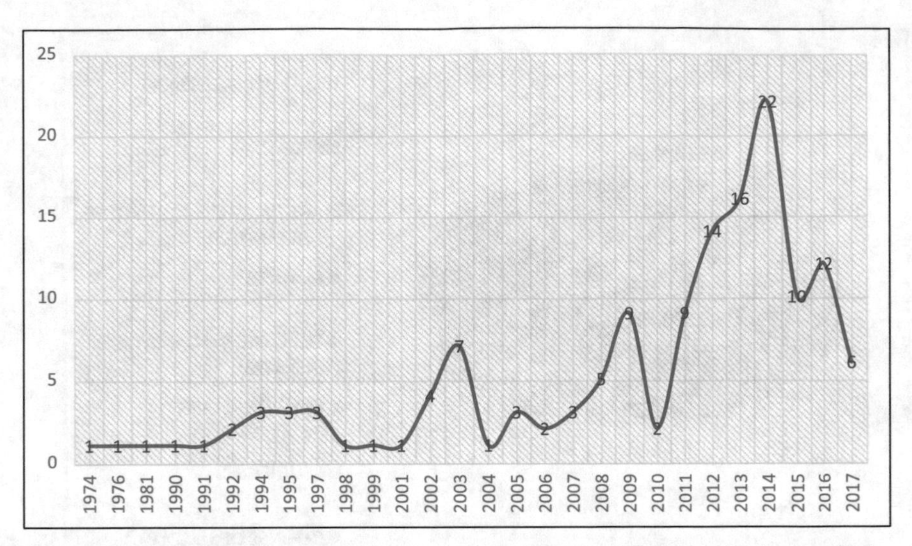

Fig. 3. Change in the number of associations created between 1974 and 2017 Source: Authors' calculations based on the ONS sample (2018).

for the children of martyrs (*Chouhada*) and veterans (*Moudjahidine*)—have joined the existing associative sphere, forming a "revolutionary family."[9]

3.5 The Associative Movement for and by Students

Various associative movements among students have shown their capacity for mobilization. Some of the student associations in our sample stand out in their dynamism, including Development House (Maison de développement), the Young Scientists' Association (Association des jeunes scientifiques), the Youth Activities Association (Association des activités des jeunes), and *Chabab bila houdoud* (Youth Without Limits). This was also the case for ACSES, which was created in 2015 with the primary objective of promoting social entrepreneurship in Algeria. This association acts as an incubator for SSE projects. It also provides students working on a solidarity project with the opportunity to start their project in a dynamic, innovation-focused environment. Training and support are provided throughout the project, up until its launch. Candidates must submit an application and defend it in front of a panel. The incubator takes on around twenty projects every year, for a period of twenty-four to thirty months. Its goal is to bring fifteen projects to fruition per year.

[9] These are organizations created informally by civil society, which are based on the values of sharing and solidarity and which work on contentious issues. In the case of Algeria, these include the descendants of martyrs and their heirs (parents of unmarried martyrs and living veterans) (Fig. 4).

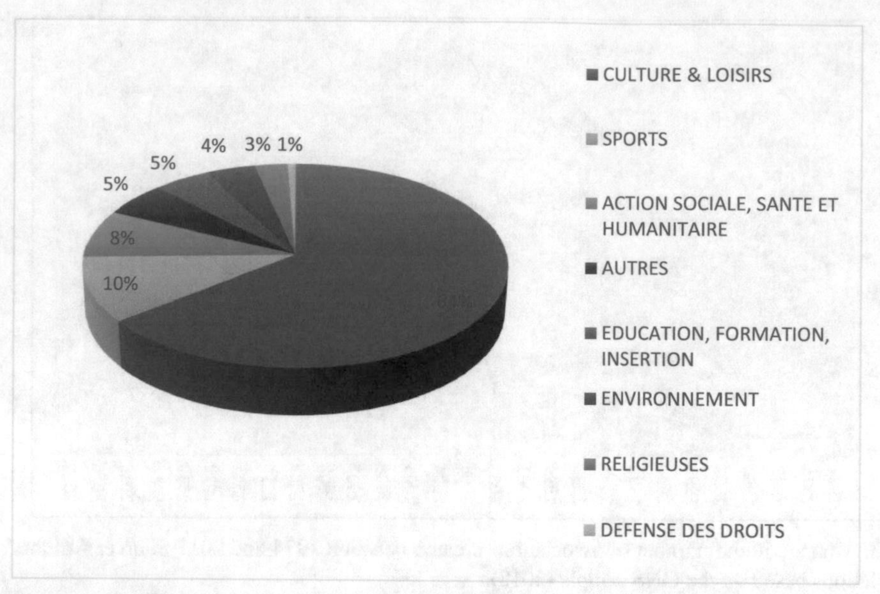

FR	EN
Culture & loisirs	Culture & leisure
Sports	Sports
Action sociale, santé et humanitaire	Social, health, and humanitarian action
Autres	Other
Éducation, formation, insertion	Education, training, integration
Environnement	Environment
Religieuse	Religious

Fig. 4. Breakdown of the associative movement by sector of activity Source: Figure created by the authors based on the ONS sample (2018).

3.6 The Associative Movement for Women's Rights

As stated above, some associations focus on defending women's rights, including the National Association for Women and Rural Development (Association nationale femme et développement rural), the Association of Women in the Green Economy (Association des femmes en économie verte), and the Association of Algerian Women (Association des femmes algériennes). As an important driver of state mobilization, the feminist associative movement has given itself the objective of fighting against fundamentalism, violence, and discrimination against women, promoting women's liberation with concrete civil society actions, such as environmental protection (Tahir Metaiche and Bendiabdellah 2016).

3.7 Obstacles

The many promises of public action to support the Algerian associative movement are far from being realized. There are many limitations that continue to block the development of this sector. The Algerian public authorities do not recognize the benefit that associations may have for the general interest, and they provide no support whatsoever to help them remain viable over the long term. Rather, the state carries out inspection procedures on a daily basis, often in an abusive manner. Administrative inquiries are carried out systematically and are one of the main factors that stop people from creating an association. There are many structures that have failed to obtain authorization: even though they follow all of the obligatory procedures, they never receive a response, or even acknowledgment that their application has been received. This practice, which has been common over the last fifteen years, allows the administration to suspend the creation process without needing to justify its refusal. In many instances, it is political and human rights associations that suffer this fate. When it comes to relationships with volunteers, the Algerian associative movement has not fully realized their value as a resource or the importance of cultivating volunteer loyalty.

Associations' business model is mostly based on subsidies. For an association to receive state financing, the local authorities need to decide whether its activities are in the general interest. However, there is no legal text that explicitly defines this idea, or that of public utility. The idea of general interest is related to the statutory objective of an association. An association is said to be "in the general interest" if, through its activities, it provides material and moral support to vulnerable populations in order to improve their living conditions and well-being.

Finally, accounting and financial management are some of the greatest difficulties for Algerian associations, even those with more experience in this area, since they are required to submit double-entry bookkeeping records.[10] The result is that the existing relationship between associations and local institutions displays a shocking lack of improvement or proper management.

4 The Associative Movement in Algeria: Searching for Democracy

The February 23, 1989 constitution freed associations from direct state supervision, opening up the path toward more autonomy. The 1990 Law on Associations reaffirmed the associative movement's emancipation from the state. According to our results, the most recent law in 2012 created an associative glut: associations popped up in every area, mobilizing every social category. However, it appears that rather than accelerating the democratization process in Algeria, the development of associative life has actually slowed it down. Based on this observation, it appears to us that Daniel Brumberg's definition of "liberalized autocracy" (2003) is the best way of analyzing the associative boom in Algeria. According to Brumberg, this term refers to a set of institutional, economic, ideological, and social factors that tend to create an environment of repression, monitoring, and partial openness, and that reflect a kind of virtual democracy in which promoting a measure of political openness for associative activities is associated with permanent state monitoring of their financing, communication, cooperation, and networking activities.

[10] Defined in opposition to "single-entry bookkeeping," double-entry bookkeeping is a method that involves recording payment operations simultaneously as a credit to one account and a debit from another.

We can conclude that the process of opening up associative activities is evolving slowly but steadily. According to our analysis, the Algerian associative movement is becoming denser, more diverse, and more dynamic. The relative openness of the associative field was considered and included in the state's 2018 list of primary objectives. Nevertheless, for this process to result in democratic action, the state itself must be more democratic. As Daho Djerbal (2012) highlights, "the existence of an associative movement is one of the prerequisites for the transition from an authoritarian regime to a democracy. Without an associative movement, there cannot be a democracy." Ultimately, the bill proposed in 2018 by the Minister of the Interior, Local Authorities, and Territorial Planning gives us a glimpse of how the associative sector may become more democratized in the future.

References

Adel, F.-Z., Guendouz, A.: La gouvernance des politiques publiques en faveur de l'artisanat en Algérie, essai d'analyse sur la longue période. Marché et Organ. **24**(3), 103–125 (2015)

Akesbi, M.: L'organisation optimale d'une structure pour un accompagnement au montage et à la création d'activités ESS efficace. In: RIUESS International Conference: Comment former à l'économie sociale et solidaire? Engagement, citoyenneté et développement, Marrakesh, Morocco (2017)

Amarouche, A.: Régime politique, société civile et économie en Algérie: une analyse institutionnaliste. Mondes en développement **159**(3), 45–57 (2012)

Ben Néfissa, S. (ed.): Pouvoirs et associations dans le monde arabe. CNRS édition, Paris (2002)

Bozzo, A., Luizard, P.-J.: Les sociétés civiles dans le monde musulman. La Découverte, Paris (2011)

Brumberg, D.: Liberalization versus democracy: understanding arab political reform. Carnegie Endowment for International Peace, Middle East Series, Working Papers, vol. 37, pp. 1–24 (2003)

Charif, M., Benmansour, A.: Le rôle de l'État dans l'économie sociale en Algérie. RECMA **321**(3), 16–19 (2011)

Cherbi, M.: Algérie. De Boeck Supérieur, Paris (2017)

Dahak, B.: L'Algérie peut-elle encore éviter le pire scénario? Lematindz.net (2014)

Derras, O.: Le Phénomène associatif en Algérie. Fondation Friedrich-Ebert, Algiers (2007)

Djerbal, D.: Le défi démocratique. NAQD – revue d'études et de critique sociale **29**(1), 5–13 (2012)

Dris-Aït Hamadouche, L.: La société civile vue à l'aune de la résilience du système politique algérien. L'Année du Maghreb **16**, 289–306 (2017)

Dris-Aït Hamadouche, L., Dris, C.: De la résilience des régimes autoritaires: la complexité algérienne. L'Année du Maghreb **8**, 279–301 (2012)

Dris-Aït Hamadouche, L.: L'Algérie face au 'printemps arabe': l'équilibre par la neutralisation des contestations. Conflu. Méditerranée **81**(2), 55–67 (2012)

Dris, C.: Élections, dumping politique et populisme: quand l'Algérie triomphe du 'printemps arabe'. L'Année du Maghreb **9**, 279–297 (2013)

Ducellier, A., Micheau, F.: Les Pays d'Islam (VIIe–XVe siècle). Hachette, Paris (2016)

Izarouken, A.: Le mouvement associatif en Algérie: histoire et réalités actuelles. Réseau des démocrates (2014)

Kadri, A.: Associations et ONG au Maghreb. Aux origines des contestations. NAQD – revue d'études et de critique sociale **29**(1), 87–117 (2012)

Liverani, A.: Civil Society in Algeria. The Political Functions of Associational Life. Routledge, London (2008)
Mérad Boudia, A., Benhassine, M.: La formation sociale algérienne précoloniale, essai d'analyse théorique. Office des publications universitaires (OPU), Algiers (1981)
Mihoubi, N.: Transformation du mouvement associatif en Algérie depuis 1989. Les voies de la professionnalisation. Ph.D. dissertation, supervised by A. Kadri, Université de Paris-8 (2014)
Mokhefi, M.: Algérie: défis intérieurs, menaces extérieures. Commentaire **151**(3), 499–506 (2015)
Ould Aoudia, J.: Croissance et réformes dans les pays arabes méditerranéens. Editions AFD, Paris (2006)
Senouci, B.: Être sociétaire à Oran, témoignage. Revue Insaniyat **8**, 165–173 (1999)
Tahir Metaiche, F., Bendiabdellah, A.: Les femmes entrepreneures en Algérie: savoir, vouloir et pouvoir! Marché et Organisations **26**(2), 219–240 (2016)
Talahite, F., Hammadache, A.: L'économie algérienne d'une crise à l'autre. Maghreb-Machrek **206**(4), 99–123 (2010)
Thieux, L.: Le secteur associatif en Algérie: la difficile émergence d'un espace de contestation politique. L'Année du Maghreb **5**, 129–144 (2009)
UNDP 2017: Human Development Report 2016. Human Development for Everyone
Volpi, F.: Stabilité et changement politique au Maghreb: positionner l'Algérie dans le contexte régional de l'après-printemps arabe. Maghreb-Machrek **221**(3), 35–46 (2014)
Zoreli, M.-A.: La régulation solidaire en Kabylie. L'exemple du village de Tifilkout. RECMA **339**(1), 86–99 (2016)

Digital Storytelling for Tertiary Education in the Era of Digitization

Construction and Evaluation of Two Experiences

Khaldoun Dia-Eddine(✉)

Zurich University for Applied Sciences, Zurich, Switzerland
diak@zhaw.ch

Abstract. Globalization is everywhere. Demographic changes are visible and take more importance and pose new challenges. Technological developments are seen as a source for solutions as it is sometimes considered source of problems.

The combination of these changes along with environmental changes is impacting all what man intends to do, it influences the societal environment, creates new demands and needs. Economic success and continuous growth are not the only key words. Sustainability, innovation and more specifically social innovation are the new key words. In order to get them it is required also to have what is today called the social entrepreneurship.

Facing all these challenges requires a good preparation. Education and innovation in education are corner stones to redeem the skills' gaps.

The intention of the study is to present the changing environment as motivation and drive for change in the society and more specifically in the education, its methods and approaches. Once the drives clear the study describes two major trends as answers and opportunities, namely the smart cities and industry 4.0. In order to have a sustainable and efficient societal chapter the demand for social entrepreneurship and social innovation became a necessity. Further the issue of lack of skills required by the new technological and entrepreneurship advances is discussed showing the impact on the new conditions for education, mainly the tertiary and vocational.

The combination of all the changes and the innovation leads to new demands in education forms as well as pedagogical approaches. The study will explain some of them before concentrating on one approach known as digital storytelling (DST).

The explanation of DST will lead to the development of a framework. The framework was applied in two different ways in classes and then analyzed based on the free and anonymous feedback of the students.

It showed the interest of the students for such teaching methods as part of a curriculum, whether for the simple case or for the more complex. The students were motivated to search more and interact with the presented topics.

The positive evaluation of the students and some thoughts of the author after this practical experience open the door for some recommendations and improvements.

K. Boussafi et al. (Eds.): MSENTS 2019, LNNS 162, pp. 16–49, 2021.
https://doi.org/10.1007/978-3-030-60933-7_2

1 Methodology and Limitation

The paper is based on an extensive literature review which will explain the theoretical bases of the study and the logic behind it. It will then set the frame for the practical part. The practical part is an application of digital storytelling for university students. It was developed and conducted by the author himself in his own classes. The applications were conducted separately in two different classes and different topics for last semester's bachelor students. The evaluation of the two experiences was done by the students attending the courses on free and anonymous basis. The students were informed to fill a questionnaire designed by the author. The comparison between the two approaches-including the students' feedbacks- will feed the final recommendations of this study.

For the time being the whole experience was applied only one time, which is not representative in term of final conclusions, but allows to advance and enhance the thoughts and outcomes in the future.

Another limitation to be mentioned is the small number of students contributing to the survey. An important limitation was the lack of time, available material and technical skills to develop the stories and make the movies. At least one experience (story II) was done on the base of existing recorded material.

2 Introduction

The Greek philosopher Heraclitus [1] said once: "the only thing that is constant is change".

As technology continues to permeate not only our personal but also our professional lives, as political, demographical, social and environmental changes are impacting us, this adage is more relevant than ever – particularly for societies, businesses and individuals struggling to adapt in an increasingly moving digital economy.

The changes mentioned above created new environment, new expectations and new demands, they created volatility, uncertainty, complexity and ambiguity known as: VUCA [2]. No society or an organization's leadership, nor their strategies are spared in today's VUCA world. Societies and organizations want stability, certainty, clarity and sustainability; hence, citizens, companies, institutions and employees have to perform -as major players- at the top of their game.

Smart societies and communities, savvy organizations understand that "the top" is not a static target – it is always shifting and evolving, driven not only by competitive pressures but also by digital transformation and societal changes. To stay at the top requires wide support of the development of citizens and employee skills, knowledge and experiences. All must be learning every day, and learning must be integral to a general societal or organizational culture.

But different types of learning are possible. Education has today to offer new responses in different manners to the societies and to the economy. New factors like technical innovation, social innovation, project management, entrepreneurship and social entrepreneurships are new fields or re-invented fields. The lack of skills demanded by these changes is obvious, it is mentioned in the paper as a pushy factor for the educational institutions. This idea is also supported by many researchers [3].

It is possible to summarize the chain from the big changes to the requirement for new educational needs through this simplified diagram:

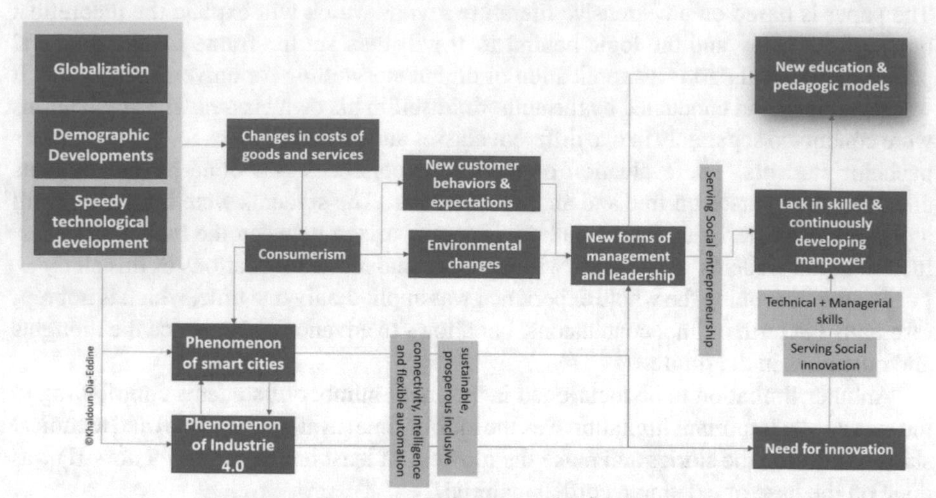

Diagram explaining the chain from changes to new educational needs

Understanding the changes, the mixes created by them, their major characteristics and their dynamics including the trends they create will help to understand the importance of innovation and entrepreneurship. But these two factors became more specific in that, that they have to answer to modern social expectations and agendas. Something new, which wasn't present for the previous generations.

As consequence, there are new needs in term of preparation for the citizens – as consumers and societal decision makers-, the workers and employees, the managers and entrepreneurs. The educational system is -up to now- not ready for these new conditions as many studies indicate[3, 4].

3 The Drives for Changes

Accelerated technological development

The fast-technological development in the last decades took several forms (among others):

The huge development in the miniaturization of integrated circuits (Moore's law) influenced dramatically the calculation and data treatment capacities [5]. This was translated in the huge increase of supercomputer power in terms of the number of floating-point operations carried out per second (FLOPS) [6] (more than 700'000 times increase between 1993 and 2017)!

The number of pages viewed from mobile devices and tablets, which is estimated to have risen from 15% to over 30% of total pages viewed in less than 2 years. In 2013, over 75% of active Facebook users connected via a mobile device. Now, more than 4 billion people have access to handheld devices that possess more computing power than the US National Aeronautics and Space Administration used to send two people to the moon [7].

Many emerging technologies rely on innovations in ICT. But that is not the only domain, in OECD countries, about 25% of ICT patents also belong to non-ICT areas.

Some of the developments were less linear than the ones just mentioned above. This non-linearity is observed where rapid evolution followed an important enabling innovation. Two examples of such trends are: the take-off of human flight, and the sequencing of the human genome (more than 175,000-fold since the completion of the first sequencing project).

The cost to keep the different machines running is also a good indicator. The progress in relation to the electrical efficiency [8] has been tremendous: researchers found that over the last six decades the energy demand for a fixed computational load halved every 18 months [9].

One interesting consequence of the technological development is the development of the price of consumer durables, which shows a decrease in the price of the majority of the large public goods (appliances, clothes, toys, etc.), this pushed for more consumption of finished products, but also of raw materials needed for the production. In contrast, the prices of goods and services such as education, childcare, medical care, and housing have increased significantly, rising up to 150% sometimes over the last 20 years [10].

This difference between price trends of consumables and services can be simply explained by the little productivity growth of service sectors relative to manufacturing sectors which have seen continued improvements through technological innovation. This may in part explain why the cost of education, healthcare and other services have risen faster than the general rate of inflation [11].

These technological changes made the ICT sector to grow such that it became one of the 10 most important economic branches in Switzerland as example [12]. The consequence is the demand for more skilled workers in this field.

The demographic changes

Demographic change and development is a two-way relationship. Demographic change affects key intermediate outcomes of development, such as the pace of economic growth per capita, savings and human capital, and these intermediate outcomes in turn influence demographic change [13]. At the same time, it is known that technological progress affects growth via the impact on labor and capital productivity [… it was observed that] demographic developments made a sizeable contribution to growth in the emerging market economies, while this contribution was much smaller or even negative in advanced economies [14].

In fact, emerging market economies tend to have more young people. This means that the share of the working age population was growing while in advanced economies it has remained either broadly stable or has declined over the past 25 years [15].

The major challenges due to the demographic changes are [16]:

- The change in the absolute size of the population (fertility dynamics and mortality dynamics). This will require new jobs and working places.
- The change in the relative size of particular cohorts of population (youth, Ageing). The change in the demographic pyramid will have its impact on retirement age and demanding for extended working life.

- The change in spatial distribution of the population (international and internal migration, concentration through higher urbanization). In this regards, rapid urban growth presents opportunities, but also challenges that seeks to make cities and human settlements inclusive, safe, resilient and sustainable (e.g. smart cities when combined with tech developments)
- The challenge related to the wealth of the population. According to some reports [17] the inequality is advancing very rapidly and the gap between the richest and the poorest on the earth is widening. This will impact the life quality, the chance for equal access to health and education services.
- Another important issue is the education. According to the World Inequality Database on Education [18], the gap in completing primary school between countries is growing. The economic productivity and social quality of life of any country depends on its educated population and closing the gap in global education is the key to global prosperity, safety and stability.

Many of these challenges must be faced through longer working life which will in its turn require continuous skills adaptation.

To conclude with the demographic aspect, Migrants made an absolute contribution to global output of roughly $6.7 trillion, or 9.4% of global GDP in 2015 [7]. This contribution could be larger if the migrants would have been better integrated through a better and more adequate education.

The challenges of globalization

Globalization, or the increased interconnectedness and interdependence of peoples and countries, is generally understood to include two inter-related elements: the opening of international borders to increasingly fast flows of goods, services, finance, people and ideas; and the changes in institutions and policies at national and international levels that facilitate or promote such flows [19].

Globalization poses challenges that will have to be addressed by governments, civil society, and other policy actors [20]. Parts of these challenges can be contained through better and continuous education.

The combination of the changes mentioned above generated new phenomenon:

1. the smart cities and
2. the industry 4.0.

In the study here both are separated from each other despite that the link between them is quite deep and organic [21]. The reason behind such separation is that the smart cities has tendency to deal -beside the technical aspects- with the societal issues parallel to the technological development. While the Industry 4.0 will -beside the technical aspects- deal mainly with the economic issues related to the economic welfare, growth along the technical development.

Trend I - Smart cities

In the last few years the development of the so-called smart cities is an important phenomenon. The importance of this phenomenon is seen in India and China [22], it is also seen through the number of connected IoT devices [23] and in the additional cities looking to be smart within the next few years [24].

But what is a smart city?

"A Smart City is a city seeking to address public issues via ICT-based solutions on the basis of a multi-stakeholder, municipally based partnership" [25].

The British Standards Institution (BSI) was even more specific when it described smart city as "an effective integration of physical, digital and human systems in the built environment to deliver a sustainable, prosperous and inclusive future for its citizens" [26].

The literature on smart cities underlines the importance of the integration of technical and social perspectives [27], but the integration is sometimes not going without creating tensions [28]. The development of human potential by means of promoting education and getting more skilled workers, can be a response to this tension and explain also why these cities have better educated people in a better endowed labor market [29]. Then the conjunction of knowledge, education and ICTs skills become key factor for analyzing all issues related to smart cities' sustainability.

One way how knowledge, education and skills are expressed within the frame of smart cities is explained through the innovation by Caragliu, et al. "..that innovation in smart city is driven by entrepreneurs and products that necessitate a progressively more capable labor force" [30].

Most of the literature defines innovation as the implementation not just of new ideas, knowledge and practices but also of improved ideas, knowledge and practices [31], as also highlighted in Oslo Manual [32].

Innovation has different types. The focus is generally much on technological innovation [33], but, the concept of social innovation has received more and more attention over the world across the last decade [34]. At the European Union's policy-making level, social innovation is already recognized as key for tackling "societal challenges," "societal demands" and "social needs" [35] in the 21st century. Realizing a shift towards sustainable societies in the end is closely linked to innovative answers and new approaches to innovation [36].

Social innovations along with technological innovations are considered as important driving force of secure sustainable development affecting wellbeing of communities, societies and countries [37, 38].

But for smart cities to be sustainable, it is not enough to have knowledge, skills and innovation. These elements must be embedded into a framework of economic activities. Hence, the presence of a high number of entrepreneurs bundled into a 'socio-technical network' [39] should allow an alignment in their offers, methods, technologies and attitudes toward smart cities and smart employees, this would lead to higher efficiency and strengthen the sustainability of smart cities.

Trend II - Industry 4.0

Besides the smart cities, the changes mentioned above, brought what is today known as the digital transformation and the industry 4.0 [40].

Industry 4.0 takes the emphasis on digital technology from recent decades to a whole new level with the help of interconnectivity through the Internet of Things (IoT), access to real-time data, and the introduction of cyber-physical systems. Industry 4.0 offers a more comprehensive, interlinked, and holistic approach to manufacturing. It connects physical with digital, and allows for better collaboration and access across departments, partners, vendors, product, and people. Industry 4.0 connects and cooperates effectively, with very close links between suppliers, manufacturers and customers [41].

Another aspect linked to industry 4.0 is the extent of automation. The proportion of jobs that can be fully automated by adapting currently demonstrated technology for middle-skill categories could rise to 15 to 20% [7]. This doesn't go without consequence on the working place. The workers of these places have to go through additional focused education processes to be able to take new tasks and activities.

Innovation is the key enabler for accelerating digital transformation and realizing the fourth digital revolution, with all the tremendous benefits it promises countries around the world [42].

Innovation doesn't happen in a vacuum, it requires openness and interactions between systems and their environments [43]. In its white paper, the world economic forum (WEF) states that the industrial revolution can only be possible if a trisector innovation system is existing comprising business, government and the social sector, including academia [44].

In the concept of Industry 4.0, IoT, IoS, IoP and IoE can be considered as an element that can create a connection of the Smart City Initiative and Industry 4.0 - Industry 4.0 can be seen as a part of smart cities [45].

To gauge the Smart technology for manufacturing process across various regions and adapt Industry 4.0 in manufacturing sector it is essential to align with the education. Educational institutions should examine multiple entry points for Industry 4.0, provide relevant education and training opportunities and expose them [46] to the necessary adaptation and conceptualize cost-effective method of adaptation that result in high productivity to expound the economies of scale.

4 The Changes, the Consumers and the Social Entrepreneurs

The "consumer" of today is the heart of the whole story. The consumer is influenced by the changes mentioned above. This same consumer can benefit from several features [47]:

The technology of intelligent voice assistants, the Augmented hearing (ex. wireless headphones), the Fem-tech revolution, where female entrepreneurs are leading the way in fem-centric businesses, the Hyper-personalized manufacturing transforming the way of shopping and the prices paid for the products, the possibility of global searching and accessibility to world markets, the trend of Airbnb-Everything-society as well as the uberization and the sharing-economy.

One can add some other characteristics [48] like Around-the-clock-shopping, control of the shopping experience, the Omnichannel shopping, the closer relationship with the purchased goods (content consumer), Global experience, the existence of "Collaborators" helping with information and the Social sharers with different feedback mechanisms.

The combination of smart cities, industry 4.0 linked to the social innovation and this new type of consumers with the new basic services and needs calls for social entrepreneurs. Social entrepreneurship is to be considered as an important factor of economic security and sustainable development [49].

Concepts like "social venturing, not-for-profit organizations adopting commercial strategies, social cooperative enterprises, and community entrepreneurship" [50] belong to the umbrella construct of social entrepreneurship, which is tasked by "the construction, evaluation and pursuit of opportunities for social change" [51]. The creation of activities which are linking innovation to social values is the common fields for the non-profit, business and public sector" [52]

Social entrepreneurs are innovators/change agents in the social sector. They have an inherent capacity to bridge different sectors and to work effectively across a variety of diverse domains or constituencies [53]. Further, they are pioneers in building "local capacities to solve problems and mobilize existing assets of marginalized groups to improve their lives" [54].

Realizing the importance of social entrepreneurship as an alternative career path in the future constellation of integrative sustainable environment; many prominent business schools, educational institutions and universities the world over has put in concerted efforts to introduce it in their curriculum [54].

The issue of technical and entrepreneurial skills
All these changes, needs and developments are not without consequences.

One of the issues in developed countries -where innovation is concentrated, and the ageing of the population is visible- is the lack of qualified and skilled labor. Rapid technological advances and the digitization of the workplace are making it harder for workers to match their skill sets with the needs of employers. The scale of the problem varies from country to country according to recent data from the OECD [55].

Switzerland's ICT Branch [56] and Germany's once [57] as well as the Austrian [58] all have the same problem: finding the needed skilled workers and employees. This is now accepted by the societies. The results of a questionnaire to the member of an education association in Wuppertaler [59] indicated that a continuous education should also be conducted within the enterprises.

The growth and the development of many companies is compromised, one third of the digitally oriented companies in Germany would like to increase the number of their employees [60].

As consequences many plans are established within the companies to cover -at least partially- this lack in skills since companies can't wait to complete their digital transformation until official and formal education generates the needed skilled personnel [60]. Many companies in Austria [58] and in Germany [57] established their own educational programs, sometimes in a non-formal form.

Parallel to that, the increased demand for social entrepreneurs became evident since the most recent figures from UK government indicates that the sector of social enterprise has a startup rate three times that of mainstream SMEs [61].

What stops the further development of the sector is the significant talent shortage at the second-line leadership and senior management levels in most social enterprises and small and medium-sized enterprises (SMEs). PricewaterhouseCoopers (PwC) concluded in a study that "Skill shortages are once again keeping CEOs awake at night, and megatrends (urbanization, changing demographics, technology) are only going to make the problem worse. This is no time for tinkering at the edges of talent management … a fundamental rethink is needed" [62].

Skills and knowledge, which can foster a societal transformation towards more sustainability, must be conveyed in lifelong learning. This must be done not only in schools (primary and secondary education) but also in higher education and adult education (e.g., in the framework of vocational education and training (VET)) [63].

5 Education for Social Transformation

Globalization, with all its strengths and drawbacks, the societal and demographic continuous changes and the phenomenon of smart cities and industries 4.0 are here to stay. As we described earlier, huge challenges in all spheres of life are to be faced. These challenges demand changes in education, not necessarily in the system or how it operates, but perhaps in its aims, and most certainly in its content [64], it requires new pedagogical models and approaches. How can the next generation of students be prepared for the social transformation they will face and be ready for the XXI century's requirements? Higher Education may have bypassed the Industrial and Taylor mass production revolutions, but they are unlikely to be as easily able to evade the very revolution they enabled through the knowledge economy [65]. The question of education's development is even more important if we know that educational systems evolve at a notoriously slow pace and this applies to their form, operation and content [64].

Today, smart societies and communities, savvy organizations understand the need for wide support of the development of citizens and employee skills, knowledge and experiences. They also understand that citizens and employees must be learning every day, and learning must be integral to a general societal or organizational culture. This integration is a process which should be adapted to new prevailing conditions and situations.

An interesting challenge to education is what several studies have found: a link between the Millennial Generations' growing use of digital tools and the distractions [66]. Many teachers believe that constant use of digital technology hampers their students' attention spans and ability to persevere in the face of challenging tasks [67]. A phenomenological study concluded that the temptation and use of social media had become a prominent aspect of university students' academic experiences [68], both within and outside of the classroom setting.

A Pearson report also shows that Generation Z's learning experiences — inside and outside of the traditional classroom — are becoming more of a routine, integral part of daily life; students are changing! This has led to students demanding changes to education, including accelerated, flexible and adaptive education options and tools [69].

At the same time, in a study, it was reported that students' cognitive and social engagement in technology-rich classrooms is significantly related to their professors' views of effective teaching [70].

Educators and parents still have the greatest impact on learning and personal development regardless of student's age [69]. The survey showed that many still value printed materials and teacher interactions as part of the college experience, but they like to have things related and linked to technology.

The conclusion of these studies -among others- indicates that technology implementation in university teaching is important, but it certainly needs to combine it with the presence of lecturer using these new technologies. This creates new needs: Faculty development programs related to changing professors' conceptions of effective teaching and use of new enabling technologies should be incorporated in the faculty preparation. There is a need to innovate the basic education.

As mentioned earlier, companies are moving into own-education programs. This is a form of further education, which allows a continuity in the qualification and makes the career of an employee more sustainable and allow the integration of work and study [60].

The key words for such continuous education are: Online-education, digital libraries, education "on demand" and interactive Modules. Here too, innovation of the education landscape within the enterprises is required.

There are important arguments to push for innovation in education: to maximize the value of public investment [71], demographic pressures, burgeoning demand for government services, higher public expectations and ever-tighter fiscal constraints. This can only be reached through innovative solutions to enhance productivity, contain costs and boost public satisfaction. In this sense, innovation could be a major driver for significant welfare gains because, educational innovations can improve learning outcomes and the quality of and equality in education provision. For example, changes in the educational system or in teaching methods can help customize the educational process [72].

This is even more needed since SDG 4 puts the provision of access to education services on the agenda pointing at common challenges across national systems: for instance, "affordable quality technical, vocational and tertiary education" (SDG 4.3) [73] and some others. Socio-economic criteria are also to be taken in that development into consideration [74].

A good example of that is already applied, for instance in the high education, where -as example- Massive Open Online Courses (MOOC) are widely used [75]. They are used for education for sustainable development [76]. MOOCs are often even considered as having a high potential for replacing former practices of learning, since they exemplify the potential of new practices in education and lifelong learning enabled by digital (hence technological) innovations.

In general, social innovation in education and lifelong learning uses technology to provide the opportunity for more appealing (e.g., playful, via gamification) or effective learning environments while fostering digital literacy at the same time. Such new practices and learning contexts would not be possible without using adequate digital devices for learning-by-doing, tailored to the intended learning effects [77].

To come to the issue of the market search for skills; education policies have traditionally focused on increasing participation in science, technology, engineering, and mathematics (STEM) disciplines. Recently, however, a more comprehensive view of innovation has emerged which recognizes the contribution of a wider set of skills and disciplines. While STEM specialists are important for certain types of innovation, particularly technological innovation, a broad view of the competencies used in the innovation and management processes has to be considered.

There are close conceptual links between innovation-specific skills and entrepreneurship skills [78]. Moreover, entrepreneurship is a critical vehicle for the introduction of innovation. During the past decade, most OECD countries have started to promote entrepreneurship skills at all levels of education [79]. Entrepreneurship education is a policy tool to develop entrepreneurial skills and encourage a more favorable culture and attitude towards innovation and the creation of new firms.

Entrepreneurship support in higher education generally has two strands. The first strand aims at developing entrepreneurial mindsets. It stresses the development of such traits as self-efficacy, creativity, risk awareness, building and managing relationships. The second strand aims to build the attitudes, skills and knowledge needed to successfully launch and grow a new business [43].

Surveys of tertiary-educated employees show that innovation requires skills allowing creativity, critical thinking and communication [43]. Skills for innovation can be grouped into three broad categories [43]:

- Subject-based skills, which represent knowledge and knowhow in a particular field.
- Thinking and creativity, including both higher-order skills and creative cognitive habits. These competencies include critical faculties, imagination and curiosity.
- Behavioral and social skills, including skills such as self-confidence, leadership and management, collaboration and persuasion.

This can only be reached through innovation in education.

In this regard and according to many researches, digital technology can facilitate [43]: Innovative pedagogic models: like gaming, Simulations, International collaborations, real-time formative assessment and skills-based assessments, E-learning aimed at autonomous learners.

This makes even more sense since digital devices and the internet have become a particularly important and powerful component of young people's lives according to OECD and other researches [80].

The bottom line of all these points is: there is a big need to develop new learning methods and tools to leverage the technological advances, to make the education part of the social innovation, hence, to be affordable, reachable and answering the market requirements for social and technical innovation. The best way to achieve modern time objectives for educational advancement is via amalgamation of technology with learning and social constructivism principles [81], knowledge is constructed through the interaction of physical, social and technological environment [82]. This can only be done when educational opportunity is created, when students and teachers engage in purposeful learning activities that help students develop in various ways. This requires clear goals, the skills to translate those goals into sound curriculum and pedagogy, and the leadership of teachers and school administrators to focus their work in supporting the creation of those opportunities [83].

6 Digital Storytelling (DST)

Story telling is one of the best strategies of teaching because it serves many purposes at one time e.g. motivating interest of learning, controlling students' behavioral problems, resistance and anxiety and building strong relationship between students and teachers [84]. Because narratives are shaped by context, storytelling is contextual [85]. Thus, students will be aware of what they have learned and the benefits that they can derive from the material [86].

With the technology advancement storytelling has become more effective in terms of functionality [87].

There have been many definitions about digital storytelling in related literature. Porter [88] defined storytelling as combining authentic stories with image, music, graphs and voice-over while Dupain and Maguire [89] described it as creating a story by integrating multimedia elements such as visuals, audio, video and animation.

Digital stories, altogether, are short videos created by integrating image, video.

background music and audible or written narration via some basic hardware and software (e.g. Microsoft Photo Story, Windows Movie Maker, Wevideo, Web 2.0 etc.) with authentic story [90].

Digital stories can have aim of informative, instructive and personal narration [91]. Digital stories can rise up in different types ranging from personal narratives or instructional stories to narratives that recount historical events and in many different fields ranging from social science to science [92].

According to Gakhar [93], digital storytelling activities allow students to increase their knowledge about the subjects they investigate. Other studies demonstrated that digital storytelling help students learn a great amount of information about real phenomena or about a curriculum [94] as well as increase their motivation and success [95].

Much has been written regarding the personal nature of digital stories and that this personal, and often emotional viewpoint is an essential element of digital storytelling, as expressed through a first-person narrative that includes a particular and personal point of view [96].

It should also be mentioned that DST is not only valid for social sciences, it is also valid and possible for the STEM branches besides the art and humanities [97].

The framework developed for this purpose consists of the following components: Idea(s) to incite wonder, main plot of the story, ideas to be learned by the students, content knowledge, human values and the moral of the story [98].

This allows us to say that DST can be used for all those looking for education in the frame of smart cities or industry 4.0 from the developers to the managers to the social entrepreneurs further to the consumers.

There are endless approaches to crafting stories, depending on purpose and audience. At least six elements are considered fundamental to digital storytelling [99]

- **Living Inside Your Story**—The perspective of each story is told in first person using your own storytelling voice to narrate the tale.
- **Unfolding Lessons Learned**—the story expresses a personal meaning or insight about how a particular event or situation touches the narrator, his/her community, or humanity.
- **Developing Creative Tension**—A good story creates intrigue or tension around a situation that is posed at the beginning of the story and resolved at the end, sometimes with an unexpected twist.
- **Economizing the Story Told**—A good story has a destination—a point to make—and seeks the shortest path to its destination. The art of shortening a story lies in

preserving the essence of the tale. This is also related to the set up of telling the story (direct, indirect, movie, life, small or larger audience, etc.).

- **Showing Not Telling**—Images, sound, and music can be used to show parts of the context, create setting, give story information, and provide emotional meaning not provided by words.
- **Developing Craftsmanship**—in communicating with images, sound, voice, color, white space, animations, design, transitions, and special effects.

To link the education today with the digital storytelling and today's students, a research [100] defined a set of 20 characteristics of the "NET-generation". These students are those students who grew up within a digital culture and have continuous access to a wide variety of technologies. The study addressed how educators can tailor their teaching strategies to match the characteristics of these learners. We added to the original table -based on the previous developments and explanations- the applicability for DST:

Learner Characteristic	Teaching strategy	Applicability to DST
Technology knowledge	Incorporate technology meaningfully	yes
Relies on search engines for information	Provide assignment that allow students to use search engines, but also critically assess the information retrieved	partially
Interested in multimedia	Includes music, videos and other media	Yes
Creates internet content	Allow students to contribute to websites, blogs, etc.	Partially
Learns by inductive discovery	Provide opportunity to students to be kinesthetic, experiential, etc.	Partially
Learns by trial and error	Allow students to test their own strategies for solving problems	Partially
Short attention span	Let students use technology to move at their own pace	Yes
Communicates visually	Allow students to use images, videos, etc. for their presentations	Partially
Emotionally open	Encourage personal interaction and self-questioning	Yes
Feels pressure to succeed	Tap students' multiple intelligences and emphasize deep learning experiences and critical thinking	Partially
Constantly seeks feedback	Provide opportunities for both positive and negative constructive feedback	Yes

How Digital Storytelling Supports Learner Characteristics and Teaching Strategies [101].

7 The Framework

Based on the above-mentioned element we would like to suggest a framework which was partially used in the later coming applications. The life cycle of a digital story is divided into four phases containing the sub-elements mentioned here below:

A- Preparation and first thoughts:

a. What are the educational objectives to reach?
b. What is the row story?
c. What are the possible and existing supporting means?
d. Who is the possible narrator? Why?
e. What is the context of the storytelling? (direct/indirect, Channel, Audience, etc.)

B- Development of the DST:
1- definition of:

a. point of view,
b. the dramatic question,
c. the emotional content,
d. the pace of the story,
e. support material.

2- The Production:

a. writing,
b. script,
c. storyboard,
d. creation of DST

C- Sharing and presenting DST
D- Evaluation of DST

a. Self-evaluation
b. Audience's evaluation, it covers the following points:

Context
Context known to me
availability of introductory information (outside the storytelling: preparation material)
story came in the right time in this course
Availability of the story at any time
Format
The used platform

(*continued*)

(*continued*)

Context
The selected format in general (combination of all elements)
The story was aligned (5)/ redundant (1) with other parts of the course under another format
Good to mix with quiz
Form
Movie
Pictures
Slides
Audio (clear)
Narrator
other persons
length (each part)
length (total)
speed of ideas and pace of story
speed of narration
transitions between ideas
transitions between parts
elements of fun
elements of surprise (of the narration)
Content
Content well structured
Content clear and understandable
Constructive and cumulative structure of knowledge
Easy and logic transitions
All elements are supporting the proposed set of information/ knowledge
Evaluation of learn effect
The level of quiz
Would evaluation be enough with quiz
were the recaps helpful for further remembering?
General evaluation of storytelling and own preferences
easyness(1 very easy- 5 very difficult)
Entertaining
Appealing
understandable/clear
amount of new information (1 too little-5 too much)

Applications:

Introduction

The wish to digitalize parts of the program at ZHAW (Zurich University for Applied Sciences) came at the same time where the author was looking for some innovations to be introduced in his courses after having got the Learning and teaching award of the School of Management and law (SML) in 2017.

Something to add, during the teaching activities the author of the study used to have guest lecturers in his modules. The author recorded most of these lectures. Prominent guest lecturers are not always available and some of the interventions were very useful and they were suitable for storytelling activities.

The basic idea

To start with first two trials of DST. The first stage was to bring two stories of different construction:

- The first one (Story I): The narrator is the lecturer (he is the author, but not the hero of the story)
- The second one (Story II): the narrator is the hero of the story.

Both stories have similar presentation frame: they are specially self-produced movie or based on self-made sequences based on a recorded lecture.

Both stories were presented in different pedagogical setup:

- Story I: in a blended pedagogical setup, where the students are not in a classroom.
- Story II: Students are following a course requiring presence in the class. During one class session the story was presented.

Each story was used in a different bachelor class, each one dealing with one different subject. The first class dealt with the topic of doing business in Middle-East, the cultural identity. The second one was for the topic War, economics and businesses, peace implementation negotiation in conflict situation

The stories in the classes

Bringing the stories to the classes happened according to the following steps:

1- Story I: within the frame of a large topic (International Business-Doing business in regionally cultural frame: The Middle East), the rules of doing business internationally were explained, the regional cultural aspects were to be added to the core knowledge. No preparation was needed, a text about the life of the poet with some of his poetries was put at the disposition of the students, but reading it was not a condition since it is not essential to understand the story or the content. Students may have had experience of other regional modules in the same course (India or China). As reminder, this session wasn't done in the classroom and it belonged to a blended learning course.

2- Story II: Theoretical base for conducting international negotiations were taught in previous sessions, models and principles were explained, which means that students were introduced to the topic. The presentation was done in the classroom of a conventional course with face to face interaction including theory, simulations and case studies.
3- For both stories, the students had to go through a quiz with some questions.
4- In the next session, students had to give their feedback about the stories in term of form and content.

Here the framework applied to both stories:

Story I:

A- Preparation and first thoughts:

a. What are the educational objectives to reach?

Explain the cultural identity of the Arabs and its elements, further it was aimed at motivating the interest of learning on strongly marked topic to overcome resistance and anxiety toward a world culture. It was also intended to allow the students a certain degree of interactivity and to advance according to their own pace. One objective was also to provide opportunities for both positive and negative constructive feedback

b. What is the raw story?
The narration is about the story of a hero who lived in the selected period of time, this personality is a very well-known Arab poet of the selected period: Zuhair bin Abi Sulma.

c. What are the possible and existing supporting means?
In order to strengthen the impact of the method, only two during-the-narration-constructed slides were used in addition to two fictive pictures of the hero.

d. Who is the possible narrator? Why?
The lecturer as the carrier of the story, he knows and carry also the knowledge as well as the values of the story.

e. What is the context of the storytelling? (direct/indirect, Channel, Audience, etc.)
A Narrator (the lecturer) tells indirectly the story of the hero. He uses the hero's life as base to explain the elements of the cultural identity. The total length was ca. 60 min cut into 7 parts of ca. 8 min each. The move from one sequence to another was only possible passing through a quiz as shown below.
The story was to be used by the student when and where they wanted and be able to repeat it as much as they wished. The evaluation of the students happens through a final exam with some questions of the content of this story.

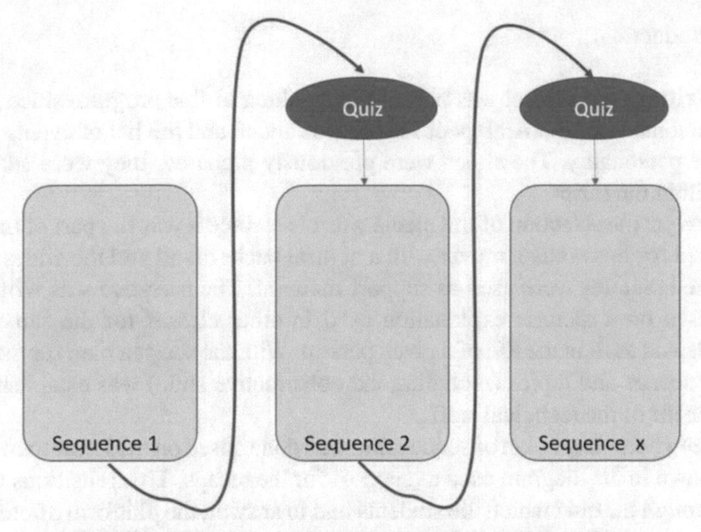

Flow of the sequences is enabled through passing the quizes, students can stop the sequence and restarts where he stopped.

B- Development of the DST:
1- Definition of:

a. Point of View (perspective): Main perspective of the story is to increase the capacity of the students to do business and interact with Arabs.
b. A Dramatic Question: The question used as guideline was what the reasons behind the Arab mind and behavior are.
c. Emotional Content: It was to transport the students into an exotic virtual trip in the time and geography with some surprising information (against given stereotypes, through some personal anecdotes of the narrator). This effect was also reached through the story of the poet and the personalization with him.
d. Use of audios and: not applicable
e. Economy: The content and effects as well as the supporting platforms were all existing elements, which didn't need large investment in time or kind. The largest time was spent to write the script, explain it to the technical staff and then the recording before the cut and montage. No frill or additional/ foreign material was used.
f. Pacing: The length and rhythm of the story were given by three elements: the time of the lesson, the time of the sequences and the number of elements to be presented in the story in order to reach the level of knowledge required.

2- The Production:

a. **Writing:** the subject was part of the teaching of this program since years. The personality of the Arab poet Zuhair was added and the list of events adapted to the personality. The slides were previously prepared, they were adapted to fit within the script.
b. **Script:** the selection of the media was clear since it was the part of the platform used for this course, movie with a neutral background and the slides plus some fixed pictures were used as support material. The narrative was written, it was based on a rhetoric explanation used in other classes for the same topic but adapted to fit in the life of a given person. With the script a plan for the sequence of scenes and topics (including the constructive slide) was established for the benefit of the technical staff.
c. **Storyboarding:** plan of sequencing was done based on the items to be explained shown in the diagram used as support for the recaps. The transitions were made through the quiz which the students had to answer, the platform offered the tools to make the transitions. No special effects (other movies, music, etc.) were used.
d. **Creating digital story:** The Moodle platform used at ZHAW, was used to support the content and the elements as well to organize the transitions and the first learning evaluation (quiz). The number of elements to be mixed was limited, it was done with the help of specialized ZHAW's staff based on clear instructions included in the script.

C- Sharing and presenting DST

Sharing digital story: The story was uploaded on Moodle and made only accessible for the students of the course. It is possible to access it wherever students has access to Moodle at any time with no specific restriction.
Screen shots

The narrator with the poet in two different poet's life periods

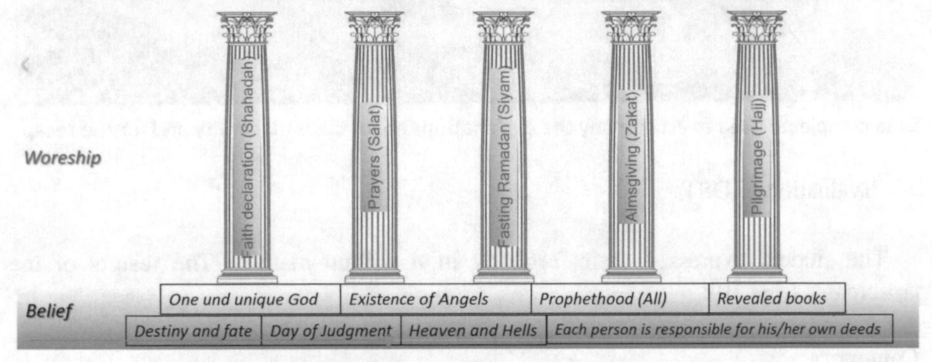

Dynamic slide used to accompany the explanations about Islam – early stage of construction

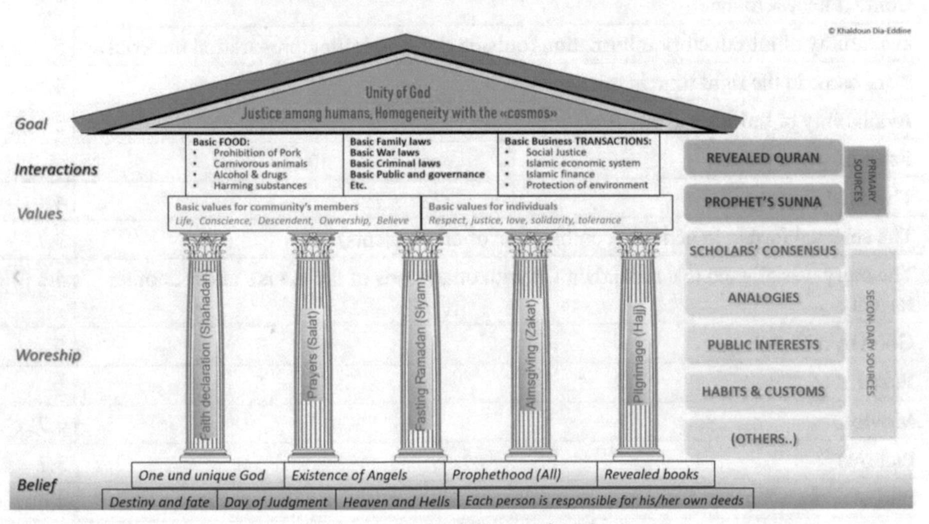

The dynamic slide completed

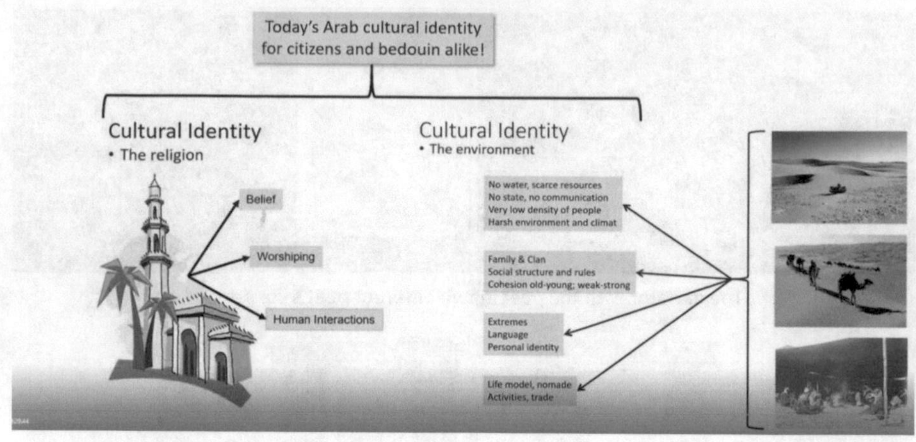

Slide completed used to accompany the explanations about cultural identity and for the recap

D- Evaluation of DST

The student expressed their feedback in a written manner. The results of the participants (n = 10):

Context	Av.
Context known to me	2.2
availability of introductory information (outside the storytelling:preparation material)	5
story came in the right time in this course	4.5
Availability of the story at any time	5
Format	
The used platform	4
The selected format in general (combination of all elements)	3.7
The story was aligned (5)/ redundant (1) with other parts of the course under another format	4.6
Good to mix with quiz	3
Form	
Movie	4.7
Pictures	3.7
Slides	4.6
Audio (clear)	5
Narrator	3.9
other persons	
length (each part)	4.9
length (total)	4.9

(*continued*)

(*continued*)

Context	Av.
speed of ideas and pace of story	5
speed of narration	4.1
transitions between ideas	4
transitions between parts	4.1
elements of fun	3
elements of surprise (of the narration)	2.8
Content	
Content well structured	4.3
Content clear and understandable	4.3
Constructive and cumulative structure of knowledge	3.9
Easy and logic transitions	4.4
All elements are supporting the proposed set of information/ knowledge	4.2
Evaluation of learn effect	
The level of quiz	2.9
Would evaluation be enough with quiz	4.7
were the recaps helpful for further remembering?	4.2
General evaluation of storytelling and own preferences	
easyness(1 very easy- 5 very difficult)	2.3
Entertaining	3.9
Appealing	4.5
understandable/clear	4.4
amount of new information (1 too little-5 too much)	4.5
changed previous stereotypes (1 too little - 5 fully yes)	3.9
Would you like to read more about this topic?	3.5
do you want more of these stories	3.7

Learning and further measures (deducted from the questionnaire including the free comments):

- Let the hero be more present (link story in a more conclusive way with the hero)
- Use some music (even traditional background music)
- Use more dialogue between the hero and other personalities in the story, introduce a second narrator a sparring partner
- Prepare a document with more text (for review and preparation for exam)
- Introduce more interaction with students (e.g. students search for old caravan routes through Arabic peninsula) → if platform doesn't allow it, make it as quiz with selection among prepared routes.
- Underline to students the importance of pre-readings → compulsory pre-reading?

- Explain the importance of Quiz in the transition → it is part of the evaluation since the program is an online course without attendance.
- It would be possible to add some complexity → n more reference to the theory of cultures

It is possible to draw some lessons from the survey. One major point is the importance of the lacks students mentioned. It is possible to come to this conclusion from the -relatively- low score for the wish to have the experience repeated. One of the objectives of this DST was to change some of the stereotypes which students had, this was widely achieved. But the DST couldn't strongly motivate the students to read more about the topic. The reasons behind that, may not be related to the DST.
Story II

A- Preparation and first thoughts:

a. What are the educational objectives to reach?
In addition to the objectives related to DST in general (mentioned in Story I), the major educational objective was to explain the development and implementation of a negotiation process in the case of conflict resolution
b. What is the row story?
Narration will inform about a person who was mandated to conduct negotiations to implement a solution according (and parallel) to international negotiations to end a civil war influenced by international powers. The narrator is the hero of the story. The story was to be used
c. What are the possible and existing supporting means?
A movie with the explanations was used in the form of recorded lecture. This was mixed with theoretical models explained earlier to the students and some additional graphs. The whole story and the supporting material were incorporated in the presentation as explained later.
d. Who is the possible narrator? Why?
A Narrator is the negotiator himself. He tells his own story. He is the best to tell these stories and events.
e. What is the context of the storytelling?
After having explained the theoretical parts related to this story, Students confront the narrator (the negotiator) who tells indirectly (to the actual students) his own story. He uses the timeline of the conflict and the rollout of the events as base to explain how the negotiation happened, the tactics used, the element influenced, the establishment and management of the relationship, the cultural hurdles, and the outcomes. Students were not introduced earlier to the conflict and its background. The total length was about one hour.

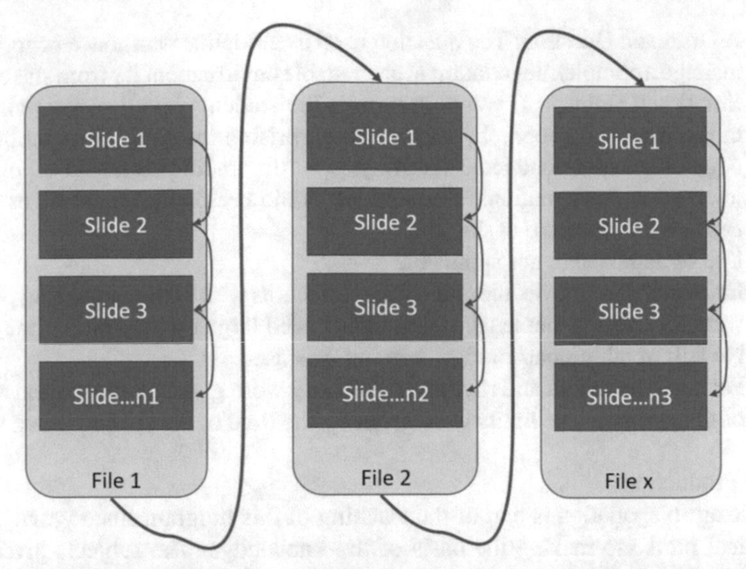

Flow of the slides and as consequence the flow of the files containing the slid
The mechanisms are not visible to the students

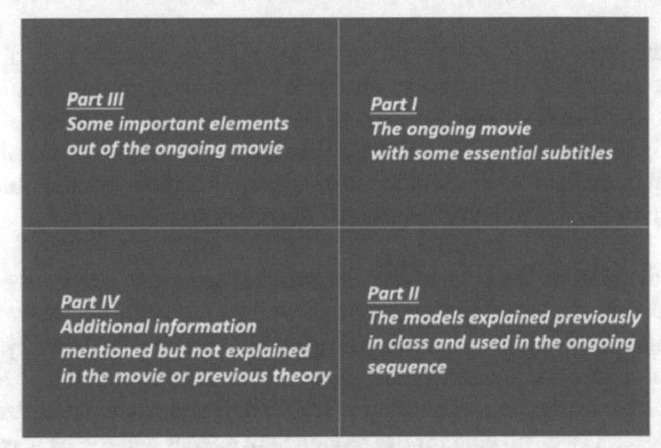

Each slide was divided into 4 parts

The story was presented to the student in the classroom once only. No evaluation of the students happened, the evaluation of this course is done through a term paper, which is not automatically linked to conflicts.

B- Development of the DST:

1- definition of:

a. Point of View (perspective): The perspective was the narrator's perspective which included his personal perspective as a negotiator and then the perspective of his employer, the UN.

b. A Dramatic Question: The question used as guideline was how a negotiator can manage a complex negotiation in an unstable environment far from stakeholders.
c. Emotional Content: It was to transport the students into an exotic virtual trip in the time and geography with some surprising information about how such negotiations are conducted far away from the traditional media reports about international negotiations. The students could personalize themselves with the narrator and the hero of the story.
d. Use of audios and: not applicable
e. Economy: The movie, the content and effects as well as the supporting platforms were all existing elements, which didn't need large investment in time or kind. No frill or additional/ foreign material was used.
f. Pacing: The length and rhythm of the story were given by the natural storyline of the narrator. The limits were given by the time of the session (max. 90 min).

2- The Production:

a. **Writing:** Negotiation is part of the teaching of this program since years. The theoretical parts are making the basis of the knowledge. The subjects given to the narrator before his intervention were: negotiation's elements, phases, types, and strategies and tactics, the conflict life cycle, issues of relationship, cultural differences, persuasion and trust. The narrator -who is the hero of the story- prepared his text.

b. **Script:** the selection of the media was done later; it was not part of the original design of the lecture or the story. The slides used were produced later as support for the story. The form and the content were not designed upfront since the idea of the "digital storytelling" came three years later. The need for the slide is due to the complex mixture of the subject: international relation, conflict and civil wars, foreign cultures, dynamics of the events, negotiation tactics, etc.

1- The division of the screen and the mix between the movie (the story) with the models explained in the theory as well as the additional information were essential to give to the students the full understanding of the story.
2- The narrative was written by the narrator based on his experience and the timeline of the events he went through. This facilitated students' understanding. In this special case, no plan for the sequence of scenes and topics was established.

c. **Storyboarding:** plan of sequencing was done based on the taken movie which was not planned to be used in this form. The transitions between the sequences were made where it was possible to make them. The transitions took into consideration the elements involved in the story, the technology used (Microsoft PowerPoint) and the nature of the given narration itself. One essential element was the length of each sequence since the size of the movies imbedded imposed limits because of the PowerPoint. The automatic transition from one slide to another and from one file to another was also quite demanding but allowed a seamless transition, which was considered as easier for the students. The timing of the appearance of the additional comments, models and explanations along the ongoing movie demanded good coordination and were thought to support the narration and enhance the knowledge outcome. No additional special effects (other movies, music, etc.) were used, it was considered that the existing elements are already enough.

d. **Creating digital story:** As tool for handling the taken movie the program MovieMaker was used to cut, trim and add comments and titles. The Windows PowerPoint was used to carry the content and the elements as well to organize the transitions. The creation of all that happened after a first evaluation of the feasibility of it based on the lecture, the objectives of the course and the available technical skills.

C- Sharing and presenting DST

After having presented it in the classroom with the presence of the course leader, the story was uploaded on Moodle and made only accessible for the students of the course. It is possible to access it wherever students have access to Moodle at any time with no specific restriction.

Screen shots

D- Evaluation of DST

The student expressed their feedback in a written manner. A form was proposed to them, the half of the students filled the form. No life discussion or exchange was possible.

The results of the participants (n = 13):

Context	Av.
Context known to me	2
availability of introductory information (outside the storytelling: preparation material)	2
story came in the right time in this course	5
Availability of the story at any time	2
Format	
The used platform	3
The selected format in general (combination of all elements)	3
The story was aligned (5)/ redundant (1) with other parts of the course under another format	5
Good to mix with quiz	n.a.
Form	
Movie	5
Pictures	n.a.
Slides	3
Audio (clear)	4
Narrator	4
other persons	n.a.
length (each part)	5
length (total)	3
speed of ideas and pace of story	4
speed of narration	3
transitions between ideas	3
transitions between parts	4
elements of fun	4
elements of surprise (of the narration)	4
Content	
Content well structured	4
Content clear and understandable	4
Constructive and cumulative structure of knowledge	4
Easy and logic transitions	4
All elements are supporting the proposed set of information/ knowledge	4
Evaluation of learn effect	
The level of quiz	3
Would evaluation be enough with quiz	5
were the recaps helpful for further remembering?	4
General evaluation of storytelling and own preferences	

(*continued*)

(*continued*)

Context	Av.
easyness(1 very easy- 5 very difficult)	4
Entertaining	4
Appealing	4
understandable/clear	4
amount of new information (1 too little-5 too much)	5
changed previous stereotypes (1 too little - 5 fully yes)	4
Would you like to read more about this topic?	4
do you want more of these stories	4

Learning and further measures (deducted from the questionnaire including the free comments):

- The shooting of the movie was not presenting the screen behind the narrator, this required additional information space on the slides, this should be avoided for other movies.
- The slides were filled with a lot of information, this reduced the visibility and presented the movie in small format.
- Having all this information on slides disrupted the students from following the narration.
- Having the whole story done at one shot was too heavy for the students and required too much attention and energy. Maybe it can be cut in order to get a rest.
- Use some music (even traditional background music from the region)
- Prepare and distribute in advance a document with additional text and information as pre-reader. This allows students for the topic and the context.
- Introduce more interaction with students through cutting the sequence of the story with one or two quiz to be discussed in the class with the students.

If we look to some other points, we may see that students benefited from the experience. They changed some of the stereotypes they had previously, they are demanding for more of such experiences. Students didn't appreciate the non-availability of the movie for additional watching for example.

8 Conclusions

This experience was certainly only a first step and not a representative one. Despite that, many lessons can be learned from this experience. One of them is to combine more elements in the DST, like music, another person, pre-reading, etc. Another possible improvement is about the length and about the transition between the sequences.

Some of the corrections made by the students should be done.

The experience shows that there is a good interest in this way of teaching and a certain degree of freedom in using it. These are two elements which are part of the requirements for an adapted education. The other one is the use of digital means as platform and the possibility to have very short sequences with an electronic interaction.

That said, it would be possible to increase the level of interaction with the students in two ways:

1. Interaction in the construction of the stories
2. Evaluation of the learning effect on students either through a test, a presentation related to the topic or through a comparison between two surveys: a survey before and another at the end of the DST.

References

1. Large, W.: Heraclitus. https://www.arasite.org/WLnew/Greeks/heraclitus.html. retrieved October 2019
2. According to www.vuca-world.org, VUCA is an acronym used by the American Military. It stands for Volatile, Uncertain, Complex and Ambiguous. It was the response of the US Army War College to the collapse of the USSR in the early 1990s (1990)
3. Jónasson, J.T.: Educational change, inertia and potential futures. Euro. J. Futures Res. **4**(1), 1–14 (2016). https://doi.org/10.1007/s40309-016-0087-z
4. Stewart, B., Khare, A., Schatz, R.: Volatility, uncertainty, complexity and ambiguity in higher education. In: Mack, O., Khare, A., Krämer, A., Burgartz, T. (eds.) Managing in a VUCA World, pp. 241–250. Springer, Cham (2016). https://doi.org/10.1007/978-3-319-16889-0_16
5. Parts of the following data are collected from: Max Roser and Hannah Ritchie Technological Progress retrieved from: https://ourworldindata.org/technological-progress
6. FLOPS) is a measure of calculations per second forfloating-point operations. Floating-point operations are needed for very large or very small real numbers, or computations that require a large dynamic range
7. McKinsey Global institute, (2017), TECHNOLOGY, JOBS, AND THE FUTURE OF WORK, Briefing note, February 2017
8. Electrical efficiency measures the computational capacity per unit of energy. This efficiency is important with respect to the environmental impact that energy production has
9. Greene, K.: MIT Technological Review (2011). https://www.technologyreview.com/s/425398/a-new-and-improved-moores-law/. retrieved 10 October 2019
10. United States Bureau of Labor statistics (BLS)
11. Max Roser and Hannah Ritchie Technological Progress. https://ourworldindata.org/technological-progress. retrieved 11 October 2019
12. Institut für Wirtschaftsstudien Basel (IWSB): ICT-Fachkräftesituation Bedarfsprognose 2024. ICT-Berufsbildung Schweiz, Bern (2016)
13. Amer Ahmed, S., et al.: Demographic Change and Development: A Global Typology, Development Prospects Group of the World Bank, Washington DC (2016)
14. Lorenzo, B.S.: Member of the Executive Board of the ECB; Demographic trends technological progress and economic growth in advanced economies; 66th National Pediatrics Congress, Rome, 22 October 2010
15. World Bank, Trends in working age population, Percentage of population aged between 15 and 64
16. UN System task team on the post-2015 UN development agenda, population dynamics (2012)

17. World inequality report (2018)
18. World Inequality Database on Education (WIDE). https://www.education-inequalities.org/ and https://borgenproject.org/closing-gap-global-education/. Retrieved November 2019
19. WHO, definition of globalization. https://www.who.int/topics/globalization/en/, November 2019
20. John, W.S.: Carnegie Council for ethics in international affairs, 1998, Challenges of Globalization Human Rights Dialogue 1.11 (Summer 1998) "Toward a "Social Foreign Policy" with Asia" (1998)
21. Anton, S., Lyudmila, K., Zoya, K.: Integration of Industry 4.0 technologies for "smart cities" development. In: IOP Conference Series Materials Science and Engineering (2019). https://doi.org/10.1088/1757-899x/497/1/012089
22. McKinsey Global Institute (MGI), Smart Cities: Digital Solutions for a more Livable Future (2018)
23. Forecast from Gartner Inc., February 2017
24. Karen, G.: Smart city market is fastest growing segment of government (2017). http://www.digitaljournal.com/tech-and-science/technology/smart-city-market-is-fastest-growing-segment-of-government/article/500081#ixzz5WlsvjhaU
25. European Union (2014), Directorate General for Internal Policies. Policy Department A: Economic and Scientific Policy. Mapping smart cities in the EU
26. The British Standards Institution, (2014). "PAS 180:2014, Smart cities – Vocabulary". February 2014
27. Levy, Y., Ellis, T.J.: A systems approach to conduct an effective literature review in support of information systems research. Inform. Sci. **9**, 181–212 (2006)
28. Joss, S., et al.: The Smart City as Global Discourse: Storylines and Critical Junctures across 27 Cities, Journal of Urban Technology (2019)
29. Glaeser, E.L., Berry, C.R.: Why Are Smart Places Getting Smarter, Rappaport Institute/Taubman Center Policy Brief, p. 2 (2006)
30. Caragliu, A., Del Bo, C., Nijkamp, P.: Smart cities in Europe. J. Urban Technol. **18**, 65–82 (2011). in Pierce, P., Andersson, B.: Challenges with smart cities initiatives – a municipal decision makers' perspective. In: Proceedings of the 50th Hawaii International Conference on System Sciences (2017)
31. Kostoff, R.N.: Stimulating innovation. In: Shavinina, L.V. (ed.) The International Handbook on Innovation, Pergamon, pp. 388–400 (2003)
32. OECD/Eurostat, Oslo Manual: Guidelines for Collecting and Interpreting Innovation Data, 3rd Edition, OECD Publishing, Paris (2005). http://dx.doi.org/10.1787/9789264013100-en
33. Godin, B.: Innovation Contested: The Idea of Innovation over the Centuries; Routledge: New York, NY, USA (2015)
34. Mulgan, G.: Social innovation: the last and next decade. In: Howaldt, J., Kaletka, C., Schröder, A., Zirngiebl, M. (eds.) Atlas of Social Innovation: New Practices for a Better Future, SI-DRIVE Deliverable No. 12.6; TU Dortmund University: Dortmund, Germany, 2018; pp. 194–197 (2018). https://www.socialinnovationatlas.net/fileadmin/PDF/einzeln/04_Future-Challengesand-Infrastructures/04_01_SI-the-last-and-next-decade_Mulgan.pdf. Retrieved on 20 December 2018
35. Bureau of European Policy Advisers; European Commission (Eds.) Empowering People, Driving Change: Social Innovation in the European Union (2010). https://www.ec.europa.eu/docsroom/documents/13402/attachments/1/translations/en/renditions/native.Retrieved on 20 December 2018
36. Wu, D., Ely, A., Fressoli, M., Van Zwanenberg, P., Bell, B., Bokor, K., Contreras, C.: New Innovation Approaches to Support the Implementation of the Sustainable Development Goals (2017). https://unctad.org/en/PublicationsLibrary/dtlstict2017d4_en.pdf. Retrieved on 20 December 2018

37. Tvaronavičienė, M.: Entrepreneurship and energy consumption patterns: case of hoseholds in selected countries. Entrepreneurship and Sustainability Issues **4**(1), 74–82 (2016)
38. Boonyachut, S.: Sustainability of community's entrepreneurship: case of floating market at Ladmayom. Entrepreneurship and Sustainability Issues **4**(2), 211–219 (2016)
39. Sauer, S.: Do smart cities produce smart entrepreneurs? J. Theor. Appl. Electron. Commerce Res. **7**(3), 63–73 (2012)
40. The First Industrial Revolution: The first industrial revolution happened between the late 1700s and early 1800s after the discovery of the steam power replacing the animal power through water and steam-powered engines and other types of machine tools. The Second Industrial Revolution: happened in the early part of the 20th century when steel and electricity were introduced in factories as well as changing the production processes (Ford). The results were increased efficiency and factory machinery became more mobile. Mass production concepts like the assembly line were introduced as a way to boost productivity. The Third Industrial Revolution: Starting in the late 1950s with manufacturers incorporating more electronic—and eventually computer—technology and some automation into their factories on the cost of analog and mechanical technology, later more digital technology and automation software were also introduced
41. Mindas, M., Slavomir, B.: Mass customization in the context of industry 4.0: implications of variety induced complexity, Industry 4.0, Mass customization, Complexity, Demand, Variety, Advanced industrial engineering, Industry 4.0, pp. 21–39 (2016)
42. ITU (2018). https://telecomworld.itu.int/2018-daily-highlights-day-3/digital-innovation-ecosystems-the-key-to-industry-4-0/
43. OECD: Innovating Education and Educating for Innovation: The Power of Digital Technologies and Skills. OECD Publishing, Paris (2016)
44. WEF and Mckinsey Company, Fourth Industrial Revolution Beacons of Technology and Innovation in Manufacturing, white paper (2019)
45. Lom, M., Pribyl, O., Svitek, M.: Industry 4.0 as a Part of Smart Cities. https://doi.org/10.1109/scsp.2016.7501015. Conference: SCSP2016 - Smart City Symposium Prague 2016 (2016)
46. European Commission, Factories of the future - Multi-annual roadmap for the contractual PPP under Horizon 2020 (2013). https://www.scribd.com/document/271903700/Factories-of-the-Future-2020-Roadmap#
47. Neuburger, H.: Consumer behaviour trends and changes to watch in the nearest future (2018). https://www.eu-startups.com/2018/04/consumer-behaviour-trends-changes-to-watch-in-the-nearest-future/. Retrieved 29 October 2019
48. Alison Bolen, SAS Insights Editor, SAS.com (n.d.)
49. Sulphey, M.M., Alkahtani, N.S.: Economic Security And Sustainability Through Social Entrepreneurship: The Current Saudi Scenario, Journal Of Security And Sustainability Issues Issn 2029–7017 (2016)
50. Hirsch, P.M., Levin, D.Z.: Umbrella advocates versus validity police: a life-cycle model. Organ. Sci. **10**(2), 199–212 (1999)
51. Roberts, D., Woods, C.: Changing the world on a shoestring: the concept of social entrepreneurship, pp. 45–51. Univ. Auckland B. Rev., Autumn (2005)
52. Austin, J., Stevenson, H., Jane Wei-Skillern, J.: Social and commercial entrepreneurship: same, different, or both?, Entrepreneurship: Theory and Pratice Journal, da Baylor University, Estados Unidos, volume 30, nmero 1, p ginas 1–22, janeiro de 006 (2006)
53. Alvord, S.H., Brown, L.D., Letts, C.W.: Social entrepreneurship: leadership that facilitates societal transformation—an exploratory study. Cambridge, Mass.: Center for Public Leadership, Kennedy School of Government, Harvard University (2003). http://citeseerx.ist.psu.edu/viewdoc/download?doi=10.1.1.321.8660&rep=rep1&type=pdf. Retrieved November 2019

54. Sulphey, M.M., Alkahtani, N.S.: Economic Security And Sustainability Through Social Entrepreneurship: The Current Saudi Scenario, Journal Of Security And Sustainability Issues Issn 2029-7017 (2016)
55. WEF: https://www.weforum.org/agenda/2016/07/countries-facing-greatest-skills-shortages/. Retrieved 11 November 2018
56. https://ictswitzerland.ch/publikationen/studien/ict-fachkraeftesituation-bedarfsprognose/ (2018). Retrieved 11 October 2019
57. https://www.bitkom.org/sites/default/files/2018-12/181213_Bitkom_Charts_PK_IT-Fachkr%C3%A4fte_final.pdf (2018). Retrieved 11 October 2019
58. https://www.bildungsspiegel.de/news/weiterbildung-bildungspolitik/1558-oesterreichische-weiterbildungsstudie-90-prozent-sehen-bildungsbedarf-bei-digitalen-kompetenzen (2018). Retrieved 11 September 2019
59. http://www.wkr-ev.de/trends16/trends16.htm (letzter Aufruf: 6.6.2017)
60. Donate Kluxen-Pyta, Bildungsbedarf für den digitalisierten Arbeitsmarkt, Analysen und Argumente, Konrad Adenauer Stiftung, Juli 2017, Ausgabe 266
61. UK Department for Digital, Culture, Media and Sport & Department for Business, Energy and Industrial Strategy, Social Enterprise: Market Trends 2017 https://assets.publishing.service.gov.uk/government/uploads/system/uploads/attachment_data/file/644266/MarketTrends2017report_final_sept2017.pdf. Retrieved November 2019
62. PWC 2014, 17th Annual Global CEO Survey: The talent challenge. https://www.pwc.com/gx/en/hr-management-services/publications/assets/ceosurvey-talent-challenge.pdf. Retrieved November 2019
63. Schrder, A., Krüger, D.: Social innovation as a driver for new educational practices: modernising, repairing and transforming the education system, Sustainability **2019**(11), 1070 (2019). https://doi.org/10.3390/su11041070
64. Jnasson, J.T.: Educational change, inertia and potential futures. Euro. J. Futures Res. **4,** 7 (2016). https://doi.org/10.1007/s40309-016-0087-z
65. Stewart, B., Khare, A., Schatz, R.: Volatility, uncertainty, complexity and ambiguity in higher education. In: Mack, O., et al. (eds.) Managing in a VUCA World, Springer, Switzerland (2016). https://doi.org/10.1007/978-3-319-16889-0_16
66. Kusnekoff, J., Munz, S., Titsworth, S.: Mobile phones in the classroom: examining the effects of texting, twitter, and message content on student learning. Commun. Educ. **64**(3), 344–365 (2015). https://doi.org/10.1080/03634523.2015.1038727
67. Richtel, M.: (2012) Technology Changing How Students Learn, Teachers Say. http://www.nytimes.com/2012/11/01/education/technology-is-changing-how-students-learn-teachers-say.html?pagewanted=all. Retrieved November 2019
68. Flanigan, A.E., Babchuk, W.A.: Social media as academic quicksand: a phenomenological study of student experiences in and out of the classroom (2015). https://doi.org/10.1016/j.lin-dif.2015.11.003
69. Pearson Global Research & Insights, (2018), Beyond Millennials: The Next Generation of Learners, https://www.pearson.com/content/dam/one-dot-com/one-dot-com/global/Files/news/news-annoucements/2018/The-Next-Generation-of-Learners_final.pdf
70. Gebre, E., Saroyan, A., Bracewell, R.: Students' engagement in technology rich classrooms and its relationship to professors' conceptions of effective teaching. Br. J. Educ. Technol. British J. Educ. Technol. 83–96 (2014)
71. Mulgan, G., Albury, D.: Innovation in the Public Sector, Strategy unit, Cabinet Office, London (2003)
72. OECD: Innovating Education and Educating for Innovation: The Power of Digital Technologies and Skills, OECD Publishing, Paris (2016)
73. Leal, W.L., Mifsud, M., Pace, P.: Handbook of Lifelong Learning for Sustainable Development; Springer International Publishing: Cham, Switzerland (2018)

74. OECD (Ed.) PISA 2015 Results (Volume I): Excellence and Equity in Education; OECD Publishing: Paris, France (2016)
75. Alumu, S., Thiagarajan, P.: Massive Open Online Courses and E-learning in Higher Education. Indian J. Sci. Technol. **54**, 9 (2016)
76. Carrera, J., Ram rez-Hern ndez, D.: Innovative education in MOOC for sustainability: learnings and motivations. Sustainability, **10**, 2990 (2018)
77. Schrder, A., Krüger, D.: Social innovation as a driver for new educational practices: modernising, repairing and transforming the education system. Sustainability, **11**, 1070 (2019). https://doi.org/10.3390/su11041070
78. OECD: Job Creation and Local Economic Development. OECD Publishing, Paris (2014)
79. Hytti,U., O'Gorman, C.: What is 'enterprise education'? an analysis of the objectives and methods of enterprise education programmes in four European countries. Education + Training, vol.46/1, pp. 11–23 (2004)
80. Kotluk, N., Kocakaya, S.: Researching and evaluating digital storytelling as a distance education tool, Turkish Online Journal of Distance Education-TOJDE January 2016 ISSN 1302-6488 Volume: 17 Number: 1 (2016)
81. Yang, Y.-T. C., Wu, W.-C. I.: Digital storytelling for enhancing student academic achievement, critical thinking, and learning motivation: a year-long experimental study. Comput. Educ. 1–47 (2011)
82. Prawat, Richard S.: Constructivisms, modern and postmodern. Educational Psychologist **31**(3–4), 215–225 (1996). https://doi.org/10.1080/00461520.1996.9653268
83. Reimers, F.M., Chung, C.K. (eds) Teaching and Learning for the Twenty-First Century, Harvard education press (2016). ISBN: 978-1-61250-923-5
84. Melanie, G.: Storytelling in Teaching, Observer, The Association for Psychological Science, vol. 17, no. 4 (2004)
85. De Fina, A., Georgakopoulou, A.: The Handbook of Narrative Analysis, Wiley (2015)
86. Dewia, N.R., Savitri, E.N., Taufiq, M., Khusniati, M.: Using science digital storytelling to increase students' cognitive ability, IOP Conference Series: Journal of Physics: Conference Series 1006 (2018)
87. Nazuk, A., et al.: Use of Digital storytelling as a teaching tool at national university of science and technology. Bull. Educ. Res. vol. 37, no. 1, 1–26 (2015)
88. Porter, B.: Digitales: The Art of Telling Digital Stories. Bernajean Porter Consulting, USA (2004)
89. Dupain, M., Maguire, L.: Digital storybook projects 101: how to create and implement digital storytelling into your curriculum. In: 21st Annual Conference on DistanceTeachingandLearning (2005). http://www.uwex.edu/disted/conference/resource_library/proceedings/05_2014.pdf. Accessed 6 June 2014
90. Wang, S., Zhan, H.: Enhancing teaching and learning with digital storytelling. Int. J. Inf. Commun. Technol. Educ. (IJICTE) **6**(2), 76–87 (2010)
91. Robin, B.: The effective uses of digital storytelling as a teaching and learning tool. In: Handbook of Research on Teaching Literacy through the Communicative and Visual Arts. Lawrence Erlbaum Associates, New York (2008)
92. Coutinho, C.: Storytelling as a strategy for integrating technologies into the curriculum: an empirical study with post-graduate teachers. In: Gibson, D., Dodge, B. (eds.) Proceedings of Society for Information Technology & Teacher EducationInternational Conference 2010. Chesapeake, VA: AACE (2010)
93. Gakhar, S.: The influence of digital storytelling experience on pre-service teacher education students' attitudes and intentions. Masters Abstracts International, **46**(1) (2007)
94. Malita, L., Martin, C.: Digital Storytelling as web passport to success in the 21st Century. Procedia – Soc. Behav. Sci. **2**(2), 3060–3064 (2010)

95. Wang, S., Zhan, H.: Enhancing teaching and learning with digital storytelling. Int. J. Inf. Commun. Technol. Educ. (IJICTE), **6**(2), 76–87 (2010)
96. Ballast, K.: Heart and voice: a digital storytelling journey (2007). http://www.nwp.org/cs/public/print/resource/2392
97. Sadik, A.: Digital storytelling: a meaningful technology-integrated approach for engaged student learning. Educ. Technol. Res. Dev. **56**, 487–506 (2008)
98. Hadzigeorgiou, Y.: Humanizing the teaching of physics through storytelling: the case of current electricity, Phys. Educ. **41**, 1 (2005)
99. Porter, B.: The Art of Digital Storytelling, Part I: Becoming 21st-Century Story Keepers. www.tech4learning.com. Retrieved October 2019
100. Berk, R.A.: Teaching strategies for the net generation. Transform. Dialogues: Teach. Learn. J. **3**(2), 1–23 (2009)
101. Robin, B.R.: The power of digital storytelling to support teaching and learning, digital education review. In: Berk, R.A.: Teaching strategies for the net generation. Transformative Dialogues: Teach. Learn. J. **3**(2), 1–23 (2009)

The REXI Method: A Tool for Exploring Representations of Innovation in Marketing-Design: Case of the *South-East Tunisian Gsours*

Ikram Hachicha[1(✉)] and Jean-Pierre Mathieu[2(✉)]

[1] Higher Institute of Business Administration of Sfax, Sfax, Tunisia
ikram.hachicha@gmail.com
[2] Management Sciences, Grenoble, France
jpmnant@gmail.com

Abstract. In the context of Marketing-design research, this contribution sets out, modestly, a new method of exploring representations of innovation: the "REXI Method". The cognitive approach proposed in this context makes it possible to identify design representations around a Tunisian heritage: the gsour of South-East Tunisia.

Keywords: Marketing-design · Innovation · REXI Method · Cognitive representations · Service · Gsour

1 Introduction

It is because they innovate that companies survive, develop, prosper, become more competitive and, moreover, more sustainable. In this context, "*the question is therefore not whether or not to innovate but rather how to achieve it successfully*" (Tidd et al. 2006). However, "*any innovation method necessarily includes shedding light on the implicit towards the explicit*" (Corsi and Neau 2011). In this context, cognitive approaches present new operating modes such as the "REXI Method", which is located upstream of an innovation process and can be used to explore representations of innovations for better "representation" of the product-service pair for companies. The objective of this work is to present, in a first step, the REXI Method and its theoretical foundations. An empirical perspective of the method was carried out, in a second step, in the context of the gsour of South-East Tunisia. Its possible applications in marketing will be stated at the last stage.

2 Theoretical Context of the REXI Method

2.1 Marketing-Design: An Innovative Synergy in Service Logic

Multidisciplinary work dealing with innovation only confirms its transversal character, thus sketching out a hybrid research field where each approach deployed contributes to

K. Boussafi et al. (Eds.): MSENTS 2019, LNNS 162, pp. 50–63, 2021.
https://doi.org/10.1007/978-3-030-60933-7_3

it and is enriched by contributions from other disciplinary fields (Le Masson et al. 2006) such is the case with "Marketing" and "design". In this crossed vision, still amalgamated (Kneebonne 2002; Holm and Johansson 2005) but quite legitimate in a context of creation or innovation (Mathieu 2006), various theoretical works and empirical methods abound and overlap, whether referring to a Marketing paradigm whether to design. By favoring a "Marketing" prism, we identify four classic fields of research, dominated by a "product" logic: industrial design, packaging design, graphic design and environmental design (Aubert-Gamet 1992; Lemoine 2009). Note that a new line of research has just enriched this theoretical landscape, namely the service design (Cova 2004; Hachicha 2013; Hachicha and Mathieu 2009a, 2009b, 2009c, 2010, 2011, 2012, 2013, 2014). This still embryonic conjunction, from a marketing point of view, is part of a paradigmatic renewal (Grönroos 2006; Vargo and Lusch 1999) which seeks the question of mythical dichotomy between "product" and "service" and offers exchange models dominated by a service logic. By considering products as "tangibilized" services (Vargo and Lusch 1999), the boundaries between the two must be blurred and redesigned in a single systemic "product-service" offer.

In this paradigmatic renewal, design, due to its potential for creativity and innovation, offers perspectives that break with a traditional marketing approach "*based on a reductive determinism of reality*" (Le Bœuf 2004). The result is a field of research: Service Marketing-design, which preaches towards a dialectic of exchange having as a point of reference a "product-service" offer logic and which favors a collaborative marketing approach (marketing with). Indeed, Marketing-design is by no means the place of predilection for the provider of the offer if we consider the evolving relationship of demand/supply from a consumer perspective (Filser and Trinquecoste 2009). The point of view developed considers, concomitantly, the context of supply and consumption through actors in an entire market. Since the human factor is closely linked to the innovation process (Temri 2000), we are aware that the integration of actors throughout the innovation process is a sine qua non for its success. If this paradigmatic shift will be able to broaden the classic theoretical scenarios, it requires methods which are specific to it and not the traditional ones adopted. It is in this context that the "REXI Method", potentially generalizable in other fields of research, could enrich, modestly; work on innovation methods in Marketing-design and this by explaining a whole methodological process having a strong foundation.

2.2 Innovating Through a Design Cognitive Approach

Beforehand, it seemed to us more opportune to reflect on our approach to "design". And it is the artificial sciences, sciences of design (conception) (Simon 2004) which constitute the backdrop of our proposal. This research prism would include all the sciences whose vocation is to study what is built by man, in other words "the artificial" as opposed to the "natural" (Simon 2004). It has to be said that in the world of man's artificial creations (Donnadieu 2004), the ingenium of Vico or the designo of Leonardo De Vinci (in other words "the design on purpose" or "the modeling") reigns more particularly; which refers, on the one hand, to the etymological bivalence of the term design and on the other hand to the (cognitive) capacity of the human mind to invent and create (Donnedieu 2004). And it is in this cognitive register that design should be considered as a cognitive representation. From a marketing point of view, we thus go beyond the classic restrictive approaches to a single design of "product" (all that is tangible) for reconsiderations of the design of any artifact, in other words "any tangible entity or not, conceived in order to meeting needs" (Micaëlli and Forest 2003). It follows that "*to imagine and produce an artifact (...) the intelligence of the designer is manifested in his capacity to project himself, to have an intentionality*" (Micaëlli and Forest 2003). We will therefore seek to mobilize cognitive representations of the various stakeholders or "actors-Designers" in Marketing-design in order to facilitate and promote the emergence of potential for innovation in this field of research.

From the point of view of cognitive psychology, creativity is not the preferred field of designers, who "*are themselves users*" (Margolin 1997). We adhere, as such, to the point of view adopted by the Simonian approach which considers that "*whoever imagines some arrangement aiming to change an existing situation into a preferred one is a designer*" (Simon 2004). Thus by "actor-designer", we mean any designer intentionally involved in the design (the representation) of the "service-product" artifact, and who can be: the supplier of the offer, the service provider, the consumer, the employee etc.

2.3 The REXI Method: From Conception to Exploration

Marketing research focused on innovation linked to human beings and not technological innovation, proposes "*a codesign, a co-development of representations (...) to bring together the two imaginaries of designers and users*" (Musso et al. 2005), but therefore ask methodological questions since the adoption of traditional marketing tools seems inadequate to explore the potential for design innovation. In this context of multi-actor innovation and in our opinion, research in Marketing-design deprives itself of the contribution of cognitive psychology and models of human memory. This is how the approach proposed in this contribution is part of a more global methodological framework, shedding light on a collaborative method centered on "actors-designers".

The Simonian question on artificiality brings new cognitive discourse that makes it possible to reconsider human thought in a complex systemic approach. This proposes a modeling of the complexity inherent in any phenomenon, artifact or complex perceived system, based on representations (Le Moigne 2006). The adoption of such a systemic approach will make it possible to understand the complexity of the product-service offer by trying to make its invariant elements more intelligible and to identify the actors who compose it, their interactions and relationships, but yet to clarify the purpose inherent in any system. Our thinking process was built, thus, in three stages:

- It is a question of extricating from the complexity of the offer (Morin 2005; Hachicha and Mathieu 2009a), through the systemic approach, a representation namely: "a design".
- And since there is no modeling without a modeler having his own projects (Le Moigne 2006), we will be in the presence of interwoven intentional cognitive representations. In other words, a system of cognitive representations which presents itself, then, as a systemic tool for exploring innovation potential, making it possible to identify the levers of optimization in Marketing-design;

– To give meaning, in depth, to the system of cognitive representations, the mobilization of models of human memory in cognitive psychology must be considered. Our proposal is thus structured around an idea that when an actor-designer represents an artifact (service, product), he will use his knowledge in memory acquired from past experiences with the service-product. And it is within this framework that the EPSS taxonomy (copies, properties, diagrams and scripts) was privileged (Mathieu 2001) such as has been realized in representations of food products by consumers (Mathieu 2002) and representations of marketing (Mathieu and Roehrich 2005).

From this methodological sequence, going from the complexity of the phenomena to be studied to the extirpation of representations via a systemic approach, their classification and their interpretation, emerges a method of exploration of innovations representations that we have designated by: "REXI Method" whose contribution can be considered in an innovation process for a better (re) positioning of the offer (service-product) in Marketing. And we propose in what follows, an empirical application of the method stated in a context having an evocative symbolism: the gsour of South-East Tunisia.

3 The Gsour of South-East Tunisia: An Example of the REXI Method Application

This typical ancestral heritage (Fig. 1) testifies, at the same time, to an intercultural mixing between an indigenous Berber population and an Arab-Muslim civilization, to an ingenious response of man to an arid and desert nature and to timelessness due to a dynamic of renewal of uses or (re) designs.

Fig. 1. Gsar Ouled Sultan

3.1 The Design of the Gsours in South-East Tunisia: A System of Complex Representations

By the design of an object, we only predict its use. In other words, it is a question of imagining and projecting forms of user-object situation (Cova 2004). We will thus be brought to account for the multiplicity of designs of the Gsourian heritage object, which can testify to an aesthetic, agricultural, social, functionalist, cultural aim, and many others (Hachicha and Mathieu 2012).

Gosurian design is conceived by both intangible (traditions, oral expressions, sociocultural practices, etc.) and tangible (architecture, materials, etc.) typical elements of the Gsour. The "*birth certificate*" (Louis 1975) of the latter dates back to the 11th century following the Hilalian invasion, a set of Bedouin tribes coming from Egypt, and, moreover, the resulting opposition between the Berbers, peoples natives, and Arab invaders. The mixing of the two distinct Arab and Berber cultures could not be done outside these buildings. Also the Gsourian design illustrates the expression of a vernacular architecture through which the Gsourian culture can be expressed (Belhassine 2008). In fact, "*The gsar is above all an attic made up of storage cells called ghorfas for the use of one or more tribes*" (Belhassine 2008). It is a bioclimatic building, built with local materials (stone, gypsum) and which blends into the natural environment of the region. Due to its architecture rich in geometric shapes, it evokes the ingenuity of the natives whose objective was far from that of aesthetics (Hachicha and Mathieu 2011). Hence, different historical accounts recognize the multiplicity of uses of gsar which is at the same time defensive, agricultural and social.

Nowadays, the majority of gsour are abandoned, demolished or crumbling. In an attempt to survive, some gsour nowadays fill a function totally foreign to the original one since there has been a redefinition of a touristic function of the gsar (Abichou et al. 2007). This is more in line with tourism of passage (guided tour, souk, museum, cafes and restaurants) and events (festival, setting for film shoots) than residential (hotel, lodge). The touristic promotion of this heritage is still in its infancy and must testify, in our opinion, of the multi-facets of gsour design.

3.2 The South-East Tunisian Gsours: A Contextual Design with a Touristic Reach

Gsourian design has been considered as a potential touristic service (Hachicha and Mathieu 2011). The latter is conceived as an architectural service of a complex nature (Chiadmi et al. 2009). In this case, how can such a complex product-service package be made intelligible and designable? This brings us back, first of all, to the original thoughts of Shoastck (1982), "*how to design a service*?". From a marketing point of view, the literature enunciates different spaces of thought for a service design ranging from traditional environmental approaches to those renewed experiential. The services design remains, to this day, a polysemous concept under construction. And it is in this context that we join, mainly, the work of Hachicha (2013) and the one of Hachicha and Mathieu (2009a, 2009b, 2009c, 2010, 2011, 2012, 2013 and 2014) who, based on the complexity of the "service" artifact, propose a systemic modeling of a service design as a "system of cognitive representations". In other words, in a tourism context, "*it is a question of considering the touristic service as a complex system which can be designed (representable) from a cognitive point of view*" (Hachicha and Mathieu 2011) that cannot be reduced to a single design. And it is in a multi-actor-designer scheme (consumers-tourists and professionals) that the exploration of these representations of innovation will make it possible to extract the most significant elements in order to ensure a singular touristic use for gsour without denature or distort them.

3.3 Research Methodology and Results

In our empirical approach, we mobilized both cognitive representations relating to the consumption sphere (consumer-tourists) and the supplier sphere (professionals). We thus conducted 103 semi-structured face-to-face interviews with a variety of foreign and Tunisian consumer-tourists (67 interviews) and professionals (36 interviews) acculturated either to design or to the issue of gsour and touristic services all around like designers, architects, private promoters, NGO etc. At this stage, it was not the statistical representativeness that counted but the richness of the statements, which is why we integrated a photographic corpus of the gsour (20 photos) during the realization of our interviews.

The analysis of the qualitative research material collected was carried out using the Alceste software. And it is the lexical analysis by context which seemed to us to be the most relevant for "*the identification of consumer representations in marketing actions and the identification of axes of innovation for certain consumers and for certain contexts*" (Mathieu 2004). The principle is, a division of the corpus in unit of elementary context (uce), it follows their classification in lexical universes described thanks to the words which are the most representative to them (in term of Khi2). For the interpretation phase, we use the EPSS taxonomy (Mathieu 2001) to structure and understand the meaning of each representation in context. The cognitive elements of this interpretation aid tool are classified into:

- Copies and Properties: in others words representations of a given object either in comparison with other objects that are close to it (copies), or by properties that define it.
- Diagrams and Scripts which are representations of uses, actions and more or less contextual situations often repeated with an object or with others deemed to be similar.

3.4 Analysis and Presentation of the Results

The number of consumers being twice as large as that of professionals, we have deliberately and separately analyzed the corpus of professionals and that of consumer-tourists. In the end, we seek to identify from the bottom up, a common area of common bifurcation of representations of Gsourian tourism innovations (Hachicha and Mathieu 2013).

To facilitate interpretation, we present, in table form, the different representations identified in each sphere as well as the EPSS classification that characterizes them. We specify that these representations, contextualized with the gsour, sometimes refer to a "service linked to design" and to a "design linked to service" allowing us to attribute a meaning to conjunctions of "service design" and not to one or the other.

3.5 The Innovation's Representations of Tourists-Consumers

The analysis revealed the existence of 6 classes, lexical universes or representations of the interviewees with 85% of elementary context units (uce) retained. The following table presents a non-exhaustive EPSS classification of the constituent elements of each representation that we have named according to the overall meaning of the words which are the most significant to it (Table 1).

Table 1. The Innovation's representations of service design in the context of gsour (consumer sphere)

The classes	Copies	Properties	Diagrams and scripts
A symbolic/ideal service design (12.25%)	Heritage (43.61), Wall (14.74), Food (14.74), Tunis (12.4), Country (10.97)	Simple (21.3), Social (14.74), Value (13.43) Original (11.31), Extraordinary (10.68), Design (10.3)	Have (70.74) Spirit (21.3) Creativity (16.17) Conserve (61.96) People (19.14) Create (19.62) Need (28.82)
A Peripheral sensory ornamental-scenographic service design (20.32%)	Music (134.51) Color (100.97) Light (93.68) Candle (47.82) Sheet (50.4) Bed (30.39) Pool (30.19) Cushion (26.45) Hotel (24.51) Dances (26.18)	Decorative (69) Folk (61.06) Traditional (43.08) Sifted (39.74) Luxury (26.08) Typical (25.37) Soft (23.71) Authentic (21.04) Comfort (19.55)	Keep (76.23) Color (100.97) Remain (59.04) Change (26.23) Introduce (22.29) modernism (37.61) Return (18.66) Animate (9.85)
An architectural/aesthetic-evasive service design (26.92%)	Ksarhaddada 7 (73.15) Ksarouleddabbeb 3 (50.08) Shape (69.33) Staircase (66.39) Floor (42.85) Construction (35.41) Vault (34.33)	Design (69.5) Original (59.66) Beautiful (50.31) Beautiful (31.75) Curious (24.26) Pretty (21.93) Inlaid (21.96) Organization (19.19) Rounded (16.82)	Climb (19.25) Give (17.06) Revisit (16.42) Build (12.25) Asking (11.98) Feel (11.91) Prove (10.92) Climb (10.92)
A basic socio-historical service design (15.34%)	Ksar haddada (32.42) Course (11.02) Coffee (10.34) Tataouine (9.37) Ruin (7.79) Chenini (7.73)	Knowledge (11.56) Accessible (7.73) Tourism (7.17) Hand-Crafted (6.91)	Visit (192.06) Guide (191.07) Local (37.87), Telling (74.4) Story (64.49) in Group (99.54) Friend (76.82) Alone (50.05)

(*continued*)

Table 1. (*continued*)

The classes	Copies	Properties	Diagrams and scripts
A historicist/symbolist service design (8.75%)	Tunisia (200) Castle (109.97) Palace (102.99) King (84.35) Ksar el jem (73.71) Matmata (57.25) Garden (16.56) Tower (12.66)	Large (22.4) Grand (6.9) Habitable (5.73) Rich (4.76) Cultural (4.54) Symbol (3.48)	Possess (213.51) Television (127.04) See (95.9), Hear (76.38), Postcard (62.83) (73.71), Know (42.12), Speak (41, 28)
An experiential/cultural service design (16.42%)	Staging (98.69), Museum (65.41) Food (61.12), Meal (56.83) Dress (37.54) Theater (34.86) Tools (25.87), Wax doll (24, 99)	Traditional (49, 16) Alive (25.3) Specific (21.19) Bass (20, 47) Equipment (19.73) Animation (19, 63) Berber (12.86) Native (11.13)	Participate (104.69) Prep (71.15) Eat (55.11) Display (35.97) Carry (31.93) Show (22, 29) Put on (14.57) Learn (12.51) Live (11, 78)

*The values in brackets represent the x2 (Khi2) statistically significant at the threshold of $\alpha = 0.10$ and a degree of freedom of 2.7. The higher the x2, the more specific the word is to the definition of the class

3.6 The Innovation's Representations of Professionals

The analysis revealed the existence of 4 classes, lexical universes or representations of the interviewees with 89% of elementary context units retained. The following table presents a non-exhaustive EPSS classification of the constituent elements of each representation (Table 2).

Table 2. The Innovation's representations of service design in the context of gsour (professional sphere)

	Copies	Properties	Diagrams and scripts
A Generic/basic service design (41.45%)	Hotel (31.6), gsar (17.87) Room (14.16), Restaurant (13.07) Guide (12.62), Tourism (11.66), Museum (10.52), Swimming pool (10, 04)	Comfort (14.49), Touristy (13.08), Charm (10.28), Modern (10.04), Safety (7.14), Luxury (7.14), Cultural (6.5), Local (6.18)	seek (14.26), Think (12.26), enhance (8.51), renovate (7.12), restore (4.99), Put (4.63), Host (4.4)

(*continued*)

Table 2. (*continued*)

	Copies	Properties	Diagrams and scripts
A Symbolist/historicist experiential service design (15.13%)	Work (67.62), gsar Haddada (61.81), Visit (55.57), Study (51.49), Tataouine (46.28), gsour (34.63), Douiret (30.5), Castle (28.35)	Fresh (11.26), Pleasant (6.25), Arabic (6.25)	Know (121.83), Remember (45.67), Keep (22.63), Possess (11.26), See (10.59)
A Sensory, cultural and experiential service design (17.31%)	Food (97.41), Music (79.13), Habit (67.64), Folklore (37.7) Meals (33.93) Campfire (29.01), Cooking (28.04), Light (25.59), Workshop (25.1), dances (24.13)	Traditional (114.28), Berber (31.37), Culinary (28), Daily (24.16), Sifted (18.91), Modern (18.26), Tunisian (14.53), Typical (14, 04)	Participate (88.11), Carry (77.13), Prepare (47.55), Eat (14.53), Live (8.25), Play (4.53), Imagine (3.48)
An Aesthetic-architectural service design (26.10%)	*Gsar haddada* (42.36), gsar ouled dabbeb (17.22), Chenini (14.32), Shape (32.09), Level (31, 41), (28.95), Staircase (26.62), Overlay (23.06)	*Design* (66.05), Original (35.83), Curious (23.06), Functional (21.65), Repetitive (20.13), Aesthetic (15.4), Beautiful (15.39)	Study (20.13), Modulate (17.22), Recall (15.92), Overlay (15.92), Love (12.91), Reside (12.72), Build (12.71)

*The values in parentheses represent the x2 (Khi2) statistically significant at the threshold of $\alpha = 0.10$ and a degree of freedom of 2.7. The higher the x2, the more specific the word is to the definition of the class

Thus, our results allowed the identification of four representations of Gsourian touristic services design from a professional point of view and six representations from a consumer-tourist point of view. We note that the quadripartite representations of service design from a "professional" perspective highlight a central representation (with 41.45% of uce,) consisting, mainly, of "hosting" copies within the meaning of Khi2 the most significant (* Hotel: 31.6). In their representations, the most important in terms of percentage of uce, namely class 2 (i.e. 151 uce classified) and class 3 (i.e. 200 uce), consumer-tourists mobilize more copies and properties as diagrams and scripts in the most significant chi2 sense.

We concede that a service design, from a marketing point of view, is defined through a common area of representations structured more by examples and properties than diagrams and scripts. However, we have observed, at a peripheral level, the emergence of potential for innovation through diagrams and scripts, in all classes and representations, whether from a professional or consumer point of view, that should be considered. Also, the innovative potential of service-related design is particularly materialized by:

- The Aesthetic-architectural services design representations defined by copies and especially "design" properties.
- The representations of scenographic services design explicitly presented from a consumer point of view and implicitly recognized from a professional point of view through some of its elements.
- The representations of ideal services design explicitly identified by consumers (class 1) but missing in the discourse of professionals, which reminds us of the words of Willems (1990), "*Humans are designers*" and supports our collaborative approach and no longer that of dominated by service providers.

3.7 Managerial Implications

On a more operational level, the REXI method has made it possible to provide an innovative touristic prism for the gsour of South-East Tunisia. The objective is to move from a myth to more adapted concrete touristic practices which draw on the imagination of real and/or potential actors for such a place of memory. Representations of touristic-cultural innovation can then be noticed at the level of existing/classic tourism concepts "(through the different scripts of renovation, preservation and Gsourian aestheticization) and those new to be developed in the Gsourian context. And it is in a multi-actor scheme that we propose a taxonomic vision of more contemporary designs having for invariant, a respect for the authentic Gsourian, for its particularities in a touristic vision (residential, passing, event) and which goes beyond mass tourism:

- ***Ideal and aesthetic-architectural service design***: this appeal to the body triggered by a particular emotion vis-à-vis the design of gsour, emerges an orientation towards sports tourism, in other words a cultural experience with a recreational and adventurous connotation for ludo-sports activities to be developed in the context gsourien.
- ***Scenographic, experiential and cultural services design***: Three innovative axes are to be considered:
  - Culinary-gastronomic tourism: following a reactive and convivial participatory research on the part of the designer-actors, it is relevant to think of a more experiential approach to Gsourian food in the restaurants of the region that already exist. But still, it is possible to appropriate the desired event character through participatory culinary shows.
  - Museum tourism: living eco-museums that highlight a dominant scenographic component that of reconstitution of the scenes, ancestral activities, the Gsourian experience through theatrical shows with real actors and a whole scenography allowing, among other things, the participation of tourists.
  - Aesthetic-artistic tourism: this is the promotion of festival/event tourism of local know-how, in other words, revisited crafts and art (weaving, construction, etc.).
- The socio-historical and symbolist services design: we project ourselves into an educational-didactic tourism where services deserve to be, either set up such as thematic training workshops, life internship; or redesigned such as the guided tour.

While being aware of the biases of such a taxonomic reflection, we assume its managerial scope with a view to a more optimized positioning (and repositioning) of service offers in the context of gsour. Moreover, it could be the subject of more in-depth studies giving rise to other more refined taxonomies which consider the socio-demographic and behavioral profile of actor-designers.

4 Perspective of Application of the REXI Method in Marketing-Design

If, in this work, the REXI Method was applied upstream of an innovation process, it opens up prospects for its deployment throughout the innovation process by presenting a methodological flexibility that can be beneficial for companies and this in terms:

- Anticipation of market developments: We thus agree with the work of Mathieu (2001, 2004, 2006) who, by exploring the knowledge of consumers by means of cognitive models, concludes that these allow them to better anticipate their uses of products/services and predict consumer behavior based on their future intention.
- Help with decision-making by shedding light on the areas of intervention in terms of innovation that may materialize at the level of the company's offer (propose new product-services/expand the portfolio of activities or innovate in the production of services), demand (change customer behavior), or even both.
- Identification of potential universes of innovation with a view to minimizing the inherent risks of failure (dissemination or reception by the market) which opens up avenues for a better market intelligence system that draws on a dynamic research approach information closely linked to the source actors.
- Identification of innovative axes for a continuous renewal of the product/service offer, already existing, with a view to a better adjustment at the same time.
- To better shed light on the structure of the market to which the marketing manager addresses himself and thus poses the possibility of a better understanding of the positioning and repositioning, in marketing, of the products and/or services. Exploring the possibilities of (re) positioning as represented by consumers in a given context, and this in a collaborative vision with market professionals, allows a certain convergence to better optimize the implementation of the marketing mix strategy.

5 Conclusion

In a context of innovation in Marketing-design, the approach proposed in this work is part of a methodological framework, shedding light on a method called "REXI Method" which is rooted in a systemic, artificial, complex and cognitive frame of reference. The result is an exploration of representations of innovation with a view to better (re) positioning the product-service offer. Although limited, our work presents a certain evolving vision of the demand-supply relationship that has been established upstream, both from a "consumer" point of view and from a "supply provider" point of view. The collaborative vision upstream materializes downstream, by an application of the method in the context of the gsour: a heritage of South-East Tunisia. From our results, the cross-checking of the cognitive representations of the actors-designers allowed to trace a common zone that of taxonomy of a Marketing-design of touristic services, contextualized to the gsour. This is how a transversal reading of the various representations through the memorial objects that define them is unraveled in a cultural Gsourian tourism (residential, passing and event tourism) which brings together authenticity, ecology, enhancement, rehabilitation.

References

Abichou, H., Sghaier, M., Valette, H.R.: La valorisation des ksours sahariens au sud-est tunisien un essai d'orientation stratégique d'un développement touristique durable. In: Actes du Colloque International «Tourisme saharien et développement durable: enjeux et approches comparatives», du 9 au 11 Novembre 2007, Tozeur, p. 609 (2007)

Aubert-Gamet, V.: Le design d'environnement commercial: un outil de gestion pour les entreprises de services. In: Actes du 2ème Séminaire international de recherche en management des activités de service, Université Aix-Marseille III, 9–12 Juin, La Londe Les Maures, pp. 2–23 (1992)

Belhassine, S.: Penser un aspect de médiation culturelle dans le sud-est de la Tunisie. In: Actes du Colloque International des Sciences de l'Information et de la Communication «Interagir et Transmettre, Informer et Communiquer: Quelles valeurs, Quelles valorisation?», 17–19 Avril 2008, Tunis, p. 35 (2008)

Chiadmi, N., Gallouj, C., Le Corroller, C.: L'innovation dans le tourisme, logiques, modèles et trajectoires. In: 14ièmes journées de recherche en marketing de Bourgogne, Dijon, 12 and 13 Novembre 2009 (2009)

Corsi, P., Neau, E.: les dynamiques de l'innovation, Hermès-Lavoisier (2011)

Cova, V.: Le design des services. Décis. Mark. **34**, 29 (2004)

Donnedieu, G.: Systémique et science des systèmes: quelques repères historiques, Association française de systémique (AFSCET), Groupe de travail: Diffusion de la systémique (2004)

Filser, M., Trinquecoste, J.F.: Faut-il réinventer le marketing. Décis. Mark. **53**, 5 (2009)

Grönroos, C.: Adopting a service logic for Marketing. Mark. Theory **6**(3), 317–333 (2006)

Hachicha, I., Mathieu, J.P.: La révolution tunisienne: un risque à opportunité en marketing-design des services touristiques gsouriens. In: Guillon, B. (ed.) 4ème ouvrage d'Oriane sur le risque, Paris, Editions L'Harmattan, Coll. Recherches en Gestion (2014)

Hachicha, I., Mathieu, J.P.: La recherche en marketing-design des services: prospective d'innovation touristique dans le sud-est tunisien. Economics & Strategic Management of Business Process (ESMB) (2013)

Hachicha, I.: Design et marketing des services: Système de représentations cognitives des innovations touristiques-culturelles dans le Sud-est Tunisien, Thèse de doctorat, IEMN-IAE Université de Nantes (2013)

Hachicha, I., Mathieu, J.-P.: Représentations cognitives des potentialités d'innovation dans le design des services touristiques autour des Gsour du sud-est tunisien. In: Morelli, P., Sghaier, M. (eds.) Communication et développement territorial en zones fragiles au Maghreb, Paris: L'Harmattan, Coll. Communication et Civilisation, pp. 229–244 (2012)

Hachicha, I., Mathieu, J.P.: Explorer des potentiels d'innovation touristique au moyen d'une méthode révolutionnaire en Marketing-Design des services: Application au cas des Gsour du Sud-est Tunisien. In: 3ième Colloque International «De la révolution du tourisme au tourisme de la révolution: Etat des lieux et stratégies de relance pour une Tunisie libre», Sidi Dhrif 12–13 Avril 2011 (2011)

Hachicha, I., Mathieu, J.P.: Le systeme des représentations cognitives du design des services: implications au contexte d'handicap. In: Meyer, V., Thiéblemont-Dollet, S. (eds.) Design des lieux et des services pour les personnes handicapées, Editions LEH Bordeaux, pp. 89–99 (2010)

Hachicha, I., Mathieu, J.P.: Le patrimoine ksourien: un bel objet de recherche pour le "Marketing-Design" touristique. In: Colloque International «Tourisme et Objectifs du Millénaire pour le Développement», Sousse 21–22 Novembre 2009 (2009c)

Hachicha, I., Mathieu, J.P.: Le design des Ksours: un potentiel de service touristique. In: Khatteli, H., Sghaïer, M. (éds.) Société en transition et développement local en zone difficile «DELZOD», Médenine: Institut des régions arides, pp. 531–543 (2009b)

Hachicha, I., Mathieu, J.P.: Le design des services: approche modélisatrice/conceptrice. In: Actes du 8ième Colloque International de la Recherche en Sciences de Gestion, Hammamet, 5–7 Mars 2009 (2009a)

Holm, L.S., Johansson, U.: Marketing and design: rivals or partners? Des. Manag. Rev. **16**(2), 36 (2005)

Kneebone, F.J.: Design et Marketing, un mariage de raison?. Revue Française du Marketing, Février **187**, 93 (2002)

Le Bœuf, C.: Proposition d'une nouvelle approche Marketing des médias. In: 1ère conférence francophone en sciences de l'information et de la communication, Faculté de journalisme de l'Université de Moscou, Mai (2004)

Le Masson, P., Weil, B., Hatchuel, A.: Les processus d'innovation: Conception innovante et croissances des entreprises. Lavoisier, Paris (2006)

Le Moigne, J.L.: La théorie du système général: Théorie de la Modélisation, Collection les Classiques du Réseau Intelligence de la Complexité RIC (2006)

Lemoine, J.F.: L'influence du design de l'environnement commercial sur le comportement du consommateur. In: Management et Sciences Sociales, l'Harmattan, Paris, sous la direction de Luc Marco, p. 55 (2009)

Louis, A.: Tunisie du Sud: Ksars et villages de crêtes. Editions du centre national de la recherche scientifique, Paris, France (1975)

Margolin, V.: Getting to know the user. Des. Stud. **18**, 227–236 (1997)

Mathieu, J.P.: Contextes, structures mémorielles et représentations: Stabilité ou instabilité des connaissances en mémoire. In: Mathieu, J.P. (ed.) Design et Marketing: Fondements et Méthodes, Paris, Collection Recherche en Gestion, l'Harmattan, vol. 435, pp. 161–177 (2006)

Mathieu, J.P.: Analyse lexicale par contexte: une méthode pertinente pour la recherche exploratoire en Marketing. Décis. Mark. **34**, 67 (2004)

Mathieu, J.P.: La représentation d'un produit: une histoire de contexte. In: Actes du 18ième congrès International de l'Association Française de Marketing, vol. 1, pp. 455–472 (2002)

Mathieu, J.P.: Structure mémorielle et segmentation comportementale: une application au marché cunicole, Thèse de doctorat en sciences de gestion, Université Pierre Mendès France, Grenoble (2001)

Mathieu, J.P., Roehrich, G.: Les trois représentations du marketing au travers de ses définitions. Revue Française du Marleting **204**, 39 (2005)

Micaëlli, J.P., Forest, J.: Artificialisme: Introduction à une théorie de conception. Presses polytechniques et universitaires romandes, Italie (2003)

Morin, E.: Introduction à la pensée complexe. Editions du Seuil, France (2005)

Musso, P., Ponthou, L., Seulliet, E.: Fabriquer le futur: L'imaginaire au service de l'innovation. Village Mondial, Pearson Education France, Paris (2005)

Shostack, G.L.: How to design a service. Eur. J. Mark. **16**(1), 49–63 (1982)

Simon, H.A.: Les sciences de l'artificiel, traduit de l'anglais par Jean-Louis Le Moigne. Collection Folio/Essais, Edition Gallimard, France (2004)

Temri, L.: Les processus d'innovation: une approche par la complexité. In: IXème Conférence Internationale de Management Stratégique AIMS (2000)

Tidd, J., Bessant, J., Pavitt, K.: Management de l'innovation: Intégration du changement technologique, commercial et organisationnel, Edition De Boek (2006)

Vargo, S.L., Lusch, R.F.: Rethinking the role of services. In: American Marketing Association. Conference Proceedings, Winter, p. 120 (1999)
Willems, R.A.: Design and science. Des. Stud. **11**(1), 43 (1990)

Social and Solidarity Economy as a Mechanism for Achieving Social and Regional Equality in Algeria

Mostefa Bousboua(✉)

University Badji Mokhtar-Annaba, Annaba, Algeria
bousseboua@gmail.com

1 Introduction

The outbreak of the Arab uprisings in 2010 led to the overthrow of long-serving leaders. Although, the political causes, the socioeconomic factors were the trigger of these uprisings. The Arab world, with its different countries, followed almost the same economic patterns. The beginning was with inward-oriented policies aimed at achieving economic independence (post-colonial policy). This economic style was soon abandoned in favour of neo-liberal economy, which led to the formation of new economic elites with immense wealth, spread corruption, and reversed all previous economic achievements. Consequently, the Arab region have witnessed from 2010 unprecedented protests against the regimes.

However, the socio-economic difficulties in Algeria, the iconic slogan "the people want to topple the regime" has been remarkably absent from the protests in 2011. The still vivid legacy of civil strife and bloodshed of the 1990s, in addition to the satisfaction of protests demands may explain the absence of this slogan, which was present in the protests of 2019.

After the end of the Cold War, Algeria has entered into neoliberal economy. Despite the Algeria's efforts to maintain its subsidy system to ensure social equality, the oil boom in Algeria in the 2000s has led to the emergence of a new wealthy class known as "oligarchy". This latter has benefited from many economic advantages such as tax exemption, the monopoly of consumable supplies and automotive industry. Besides acquiring colossal wealth, it has become an influential part of the Algerian regime. The gloomy political and economy situation after the steady decline of oil prices deepened by the social injustice has sparked unprecedented protests in Algeria in 2019.

This study aims to shed light on a new type of economy, so called "social and solidarity economy" and its role in overcoming social inequality and regional disparity in Algeria.

2 Forms of Social Injustice in Algeria

The slogan of social justice was one of the most important slogans raised by Algeria after independence. Therefore, the state has taken a set of measures to achieve social justice,

K. Boussafi et al. (Eds.): MSENTS 2019, LNNS 162, pp. 64–75, 2021.
https://doi.org/10.1007/978-3-030-60933-7_4

through a subsidy system that was the hallmark of socialism in Algeria and preserved in Algerian version of neo liberal economy. Ironically, this system has deepened the wealth inequality in Algeria.

2.1 Informal Economy

Since its independence in 1962, Algeria has relied on a central planning system through which it aimed to build up an industrial and agriculture base to ensure self-sufficiency. However, the 1986 oil crisis revealed the flaws and shortcomings of this system. The debt-to-GDP ratio has exceeded 80% in 1987 and in the face of the growing anger at regime's policy thousands of Algerian have taken to the streets in Algiers and other big cities. The army declared a state of emergency and resorted to violence to curb the protests. After five days, up to 500 young men had been killed, with hundreds more arrested in the worst bout of violence seen since independence (Allouche 2016).

Algeria began a transition from a command economy to a free-market system, the country had to undergo a tough stabilization and structural adjustment process, and it was tracking some economic reforms such as cutting costly subsidies, interest rate liberalization and privatization of state-owned companies. Consequently, thousands of bankrupted state-owned companies were sold for a symbolic price (Kahal 2018). Algeria's trade agreements with the EU had played a major role in increasing its involvement in the neoliberal economic mechanisms. These agreements were not limited to the economic aspect, but also extended to reform other sectors such as justice, education and higher education.

The distinguishing features of the introduction of capitalism in Algeria were political instability (civil war 1992-1999), disarray, incoherent transformation, corruption and concentration of the means of production in the hands of a few. That led to the emergence of an oligarchy class that was a key player in the longest serving president, Abdelaziz Bouteflika, benefited from five-year development plan, which in the period between 2001 and 2019 amounted to nearly 581 $ billion (Massaai 2010: 147).

The emergence of the oligarchy has not only contributed to the emergence of a new class that played a major role in the political and economic scene, but it strengthened the role of the informal economy. An economy of various names such as the underground economy, black economy, the shadow economy and the invisible economy. Generally, it refers to all currently unregistered economic activities that would contribute to the officially calculated (or observed) Gross National Product if observed. It can be also defined "market-based production of goods and services, whether legal or illegal, that escapes detection in the official estimates of GDP" (Smith 1994: 18). One of the broadest definitions includes "those economic activities and the income derived from them that circumvent government regulation, taxation or observation") Buehn et al. 2009: 07).

The parallel market for goods and services, the parallel exchange market and the parallel currency market are the main sectors of informal economy in Algeria. Whereas the differences in prices between Algeria and its neighbouring countries is the reason behind the flourishing of border economy, the black market exchange premium encourage this commerce, which reached 2 billion dollars according to the latest World Bank report. As for the labour market, the number of workers in this sector has reached three million workers; one can notice that this phenomenon has spread to fields that require high training. 18% of dentists, 16% of Architects, 15% of IT engineers, 14% of lawyers, 9% of accountants, 5% of professors, and 4% of health professionals are working in informal sector, an unofficial report says.

There are multiple problems with the private sector, which tries to reduce production costs by avoid having to meet certain legal obligation such as social insurance, the minimum wage and healthy work environment. A study conducted by the National Bureau of Statistics in 2016 shows that the average wage rate reaches 55.700 Algerian dinars per month in the public sector, compared to 32,600 dinars in the national private sector (APS 2018).

2.2 Tax Injustice

Distributive justice and tax equity are considering as the main pillars of social justice. Theoretically, achieving tax equity is possible through laws that encourage progressive tax rather than regressive tax, but the absence of mechanism of achieving such a purpose will deepen the social injustice, which is the case of Algeria. Since its independence, Algeria has tried to achieve a tax equity through different fiscal laws, which have witnessed many reforms to fit new challenges results from the transformation from command economy to free economy. Although, the improvement in the fiscal law, the achievement of tax justice remains difficult to achieve. According to official reports, income taxes has increased by 17.05%, the income tax on salary represents 79.33% of the total taxes, whereas the other kind of taxes (turnover tax, corporate income tax and branch tax) represent only 20.66%. In face of this predicament, the Algerian government introduced a bill in the parliament to impose wealth tax, which aims at bridging this fiscal gap. However, this low did not come to force because of pressure exerted on the government by the oligarchy. (Mokhtar 2016: 126).

The oligarchy take advantage of the legal loopholes contained in the tax law in order to practice tax evasion. The tax exemptions approved by successive Algerian governments that reach during the period 2010–2014 an estimated amount of 747.508 billion Algerian dinars, was aimed to encourage investment, however these exemptions was a cover for Tax avoidance/ evasion. According to the General Union of Algerian Traders and Craftsmen, the tax evasion reached 200 billion Algerian dinars annually, which represent 7.51% of GDP (Mikaoui 2015: 230).

2.3 Government Subsidies

Generally, the subsidy refers to all government measurement aims at keeping prices below the market level or keep prices for producers above the market level, or that reduce costs for consumers and producers whether directly or indirectly. (Luca 2009: 4) In Algeria subsidies is the symbol of the regime's plans for maintaining social peace.

Hence, the beginning of Abdelmadjid Tebboune's first term characterized with a series of promises confirm that subsidies remain untouched. In the same sense, the Minister of commerce confirm that he will take all necessary measurement to end the speculation that drives up milk prices. The price of this widely consumed product in Algeria is fixed by a decree since 2001, it costing 25 AD (around 0.35 dollar). This has allowed total consumption to grow to some 3 billion litters, of which 1 billion litters are imported (Oxford Business Group 2007, 152). Nevertheless, this milk subsidy allowed an Algerian businessman to pocket up to $1.2 million by importing freeze-dried milk between 2007 and 2009. A Panama Paper revealed. (Maurisse 2018).

In its annual budget for 2020, the Algerian government has allocated 560 billion dinar to the health sector, it is considered as the fourth most expensive item (after education, defense and local affaires). Although, the annual allocation to improve this sector, whether at the level of construction or pharmaceutical industry, it still suffering of many problems at the level of health facilities, such as high rate of occupancy of hospital bed which exceed 100% in CHUs (Teaching hospitals), the high number of daily attendances, long hours of working hours for doctors and nurses. Moreover, the availability of midicines especially cancer midicines seems to create a big problem for the government in the coming months, which is unable to continue in the policy of importing highly-cost cancer drugs because of the steady decline of oil prices.

The first beneficiary of the investment in the sector of health that is planned to worth 20 billion dollar between 2009 and 2025 (Oxford Business Group 2010, 212) was not the citizen. The upgrade and improvement of existing health infrastructure as well as construct new facilities, purchase of drugs and medical devices has been a subject to nepotism, cronyism and bribes, the corruption issues in the health sector have even been affected access to health care facilities and medication, which are subject to the same practices.

3 Development Models and Regional Disparity

It is difficult to speak of "territory" in Algeria without referring to the colonial period. The French colonization was established in Algeria mainly for economic purposes. Thus, France has sought to destroy socio-economic structures based on agriculture and local traditional industry by reshaping the urban fabric of Algeria through the impoverishment of the land for the benefit of the city, which has been rebuilt according to a European model. That created a distinction between a European space and a non-European space. Although the end of colonial period, these European spaces (the cities inherited from colonialism) are perceived as a reference for development in Algeria and have often accentuated and deepened regional inequality in Algeria.

3.1 The Relationship Between City and Countryside

The Algerian countryside formed the basis of the Algerian economy during the pre-colonial period. Most Algerians depended on agriculture; which explains the distribution of the population at that time. Indeed, less than 5% of Algerians lived in cities, and rely on traditional industries to make their living. Later, French colonization destroyed the socio-economic structures in the Algerian countryside, through the expropriation of agricultural land, the decline of traditional local cultures and the introduction of modern cultures and mechanization, etc. (Altayeb 2008: 33). These measures will encourage the tribes, whose attachment to the land was sacred, to show resistance against the occupation, knowing that the tribe was the first resistance cell against colonization in its early years. This explains why the colonial authorities dispersed the Algerian tribal fabric by creating a new urban space called "the Douar" in 1863 (Zouzou 2008: 141). These colonial policies pushed the Algerians to retreat to mountainous and rural areas, especially in areas where agricultural land is infertile. Some, however, preferred to move closer to the city to work by building slums around cities, hence the appearance of "slums" dates back to 1926-1930. Mainly concentrated around the capital Algiers, there were eighteen of them and inhabited by more than 5,000 Algerians (Kalthoum 2012: 267).

The representations of "Douars", in the Algerian collective imagination (family separation and persecution), explains the high rate of urbanisation in Algeria after independence that reached 30%. The urbanisation increased from 3 million in 1959 to more than 4 million in 196 - a significant number compared to the Algerian population the was estimated at 10 million people (Stora 1995: 33). This represented a burden for urban area, which had thereby reproduced the same colonial model, with large cities mostly surrounded by a network of small slums. The security situation that affected rural areas in Algeria during the last decade of the 20th century has considerably increased migration to the city.

Although the improvement of security situation in Algeria, and despite the efforts made by the State to reduce the phenomenon of rural exodus, World Bank figures indicate that it has worsened. The figures indicate the decline in the proportion of rural residents in the total population from 40.88% in 1999 to 27.95% in 2017 (World Bank 2019). Indeed, the lack of employment opportunities and services continues to explain this phenomenon. The national employability rate at the level of rural municipalities also shows that the employment rate in rural areas is 11% for services, 14% for public works, 16% for construction and industry, 21% for administration, 16% for other sectors and 38% for agriculture (Zouzou 2008: 215–216).

3.2 The Relationship Between Colonial Cities and Post-colonial Cities

Algeria inherited from French colonization an important infrastructure in the big cities founded by French colonization like Constantine, Oran and obviously Algiers the capital. However, due to the colonial capitalist system, France has shown great regional disparity by creating developed and economically integrated areas, concentrated in the coastal plains and inland basins but neglecting other underdeveloped areas located in the mountains, border areas and highlands. This regional and social duality was codified in the field of administrative management, so that developed areas were dominated by the law of fully empowered municipalities, inhabited by the French population, while poor and underdeveloped areas were subject to another law, that of mixed municipalities, comprising a predominantly Muslim population or what so called " **les indigènes**" (native). This situation has given rise to an unfairly distributed development system with dense basic structures in the areas inhabited by French settlers and little, if any, in the remaining areas inhabited by "**les indigènes** "(native) (Akaba 2010: 119).

The main challenge for the post-independence state has been to eliminate this regional duplication. However, the way he proceeded mainly concerned the area of administrative management. The administrative division of Algeria was reconsidered in December 1965 when it was decided to reduce the number of municipal and wilayas in Algeria to 15 wilayas (region), 91 daïras(sub-region) and 676 communes. In 1974, the number of wilayas rose to 31, that of the daïras to 160 and that of the communes to 704. In 1984, another adjustment was made to the administrative map of Algeria, bringing the number of wilayas to 48, communes in 1541 and daïras in 742 (Labichi and Alkama 2010: 109). However, linking urbanization to administrative procedures meant the creation of administrative entities without deepening the vision of the city as a regular group within social and institutional, economic, political and cultural interactions in a specific geographic area (Miri 2004: 60). This resulted in the creation of administrative entities that are not up to the towns left by French colonization. Indeed, there was therefore a distinction between cities built according to the European style, inherited from colonization and cities built after independence, which had adopted neither the same architectural style nor the same colonial infrastructure.

The conflicting relationship between the North and the South.

Understanding the relationship between the Algerian North and South also requires a return to the colonial era when the South did not interest French colonists at all. France has even tried to deport the Algerians to the "desert sands" of the South in order to monopolize the fertile North. France therefore considered the Algerian desert as a military region, where it gathered the largest French military bases in Africa by making it a test center for French nuclear bombs. But the discovery of gas and oil in the Algerian desert in 1956 reversed the situation and the Sahara then became the center of interest of the French colonists who were then ready to abandon the North for the Algerian South (Um Kalthoum 2013: 62).

This situation of the South as a periphery of the core (North), would have been completely different, if oil had been discovered in the Sahara earlier. France would have build urban area linked to the exploitation of oil wealth, especially since the colonial regime resorted to the construction of prosperous cities for Europeans, called "mining boom towns" following the mining boom in the south of Tunisia.".

Years after the end of French colonization, the nature of the relationship between the North and the South remains the same; a relationship similar to that, which exists between the core and the periphery. The Algerian South, whose area constitutes 72% of Algeria and three of its Wilayas that provides more than 92% of the country's income, still suffers from the lack of basic services, the low level of infrastructure and unemployment. Rate of unemployment exceeds 30% among young people (Marwan 2018).

Thus, the failure of post-colonial urbanization has reproduced the same problems, caused by colonial urbanization based on deliberate discrimination between European and non-European space.

4 The Impact of Social Inequality and Regional Disparity on the Socio-economic Scene

The consequences of social and regional inequality in Algeria can be observed at three levels:

4.1 Legalization of Corruption

The absence of distributive justice in Algeria, whether at the level of social classes or regions, has exacerbated the phenomenon of corruption which is now considered a legitimate means of (re) distributing income and affects the various segments of the society. In its report, the Interior Ministry said that 612 mayors out of 1,542 are involved in corruption and 1,174 local elected officials involved in suspicious cases, mainly related to the transfer of state property and in particular real estate. According to the report of Transparency International Organization "corruption perception index", Algeria is ranked 112th out of 180 countries. As for the 2017-2018 international competitiveness report, it indicates that Algeria occupies the 92nd place out of 137 countries for the indicator "irregular payments and bribes" (Belamri 2006: 3).

4.2 The Instability of the Social Scene

The lack of social justice has led to many protests in Algeria. Indeed, according to a report prepared by the Algerian League for the Defense of Human Rights, the number of protests 14,000 demonstrations per year (Boutelaji 2016), and it should be noted that the South of Algeria has become one regions where protests are most repeated. It is a fact that has troubled authority when the former Prime Minister Ahmed Ouyahia qualified the demonstrators in the South as "disruptive", mainly because the region was traditionally considered as a backup for the political regime and constituted an electoral base on which he relied on, to increase voter turnout. It should be noted that protests in the South have started to take political dimension that is neither customary nor similar to previous social protests. Indeed, the slogans of the demonstrators now denounce corruption, inequality and administrative mismanagement.

4.3 Between Demographic Imbalance and Inflation of the Administrative Apparatus

The "Core-Periphery" policy in terms of development, has led to a major demographic imbalance, with more than 3 million Algerians living in Algiers (the capital,) according to 2015 statistics. It is the smallest but the most favoured of the Wilayas of Algeria in terms of access to social services such as health, education, access to relatively advanced infrastructure and the most economically developed. On the other hand, 10% of the Algerian population lives in the South, a region which represents 80% of the surface of the country but which lacks services and amenities. As for the majority of the population (more than 36 million inhabitants), it occupies less than 20% of the country (World Population Review 2019). With the growing awareness of citizens about the development equation in Algeria. They realise that the social and economic development of their families depends on proximity to the Core, represented by the major wilayas or the chief towns of wilayas, the promotion of certain municipalities and daïras as delegated wilayas is become one of the main demands of certain protest movements.

5 The Solidarity Economy as a Mechanism to Overcome Social and Regional Inequality

The effects of the neoliberal economy impose the adoption of a critical discourse that stands out in its vision of the economy and this by giving the economy a social dimension, because the absence of social project as much as the lack of social efficiency and state control lead to the results and imbalances previously described in this study.

5.1 Definition of the Solidarity Economy

Solidarity economy constitutes a reaction to the neoliberal system valuing individual tendencies and aims mainly to establish an alternative economic model to neo-liberal regime (see Table 1). The European Standing Conference of Cooperatives, Mutual societies, Associations and Foundations (CEP -CMAF), was founded with the aim of strengthening the role of social economy actors in Europe, which has contributed significantly to the codification of solidarity economy in Europe (International Labour Organization 2013: 54). This economic model takes many names such as the social and popular economy, the economy of the third sector, the cooperative and participative economy, etc. However, as multiple as its labels and definitions are, this type of economy is aimed to satisfy human needs so that man becomes in the heart of the economic activity rather than seeking only to profit maximization.

Table 1. Differences between the neoliberal economy and the solidarity economy.

	Solidarity economy.	Neoliberal economy
The logic that governs the economic model	Homo sociologus logic	Homo economicus logic
referent object	Human capital	Physical capital
Objective	Meeting human needs	Rate of return
Means	Solidarity	Competition
Areas	Economic, social, cultural.	Economic
Economic entities	- Individuals, families that constitute a legal person. - Economic institutions seeking to reduce social inequalities. - solidarity-based entities (cooperatives, **Mutual**)	Economic institutions
Nature of economic projects	A project managed by a group for social purposes.	A project led by a person for his own benefit.
Decision-making method	Collective	Often individual

This economy type does not rely on homo economicus logic, in which men tried to maximize it profits according to cost-benefit calculation. Rather, it relies on homo sociologus logic that encourage men to take on consideration the appropriateness consequential logic that based on social cooperation. Thus, the solidarity economy involves all economic activities carried out by a specific type of business, mainly cooperatives and associations. The principles of which are characterized by targeting their objectives to serve the individual, the members or the group rather than target profit, autonomy of management, the democratic decision-making, the primacy of individuals and work over capital when income is shared (El Hedi 2015: 60).

Alliance 21 - an alliance which has contributed to the development of the rules of a solidarity economy - defines this type of economy as encompassing all activities linked to production, distribution and consumption which in turn contribute to the democratization of the economy based on citizen participation at local and global levels. It covers various forms of organization that people use to create their own means or to obtain quality products and services, through the dynamics of reciprocity and solidarity that binds private and public interests (Allard and Davidson 2008: 6). This economic model depends on social solidarity units and in so doing, citizens are not only consumers but they also participate in decision-making and stimulate growth in their countries.

Globally, the most commonly used definition of the solidarity economy is provided by Alliance 21, the group which convened the Workgroup on the Solidarity Socioeconomy: "Solidarity economy designates all production, distribution and consumption activities that contribute to the democratization of the economy based on citizen commitments both at a local and global level. It is carried out in various forms, in all continents. It covers different forms of organization that the population uses to create its own means of work have access to qualitative goods and services, in a dynamics of reciprocity and solidarity which links individual interests to collective interests (Allard and Davidson 2008: 6).

The solidarity economy is based on fundamental values such as participatory action, solidarity and the value attributed to human capital rather than physical capital. It has other aspirations including the fight against social exclusion and the defense of the common identity of society.

This economy is therefore based on independent structures, which can vary from one society to another. It is based on participatory democracy as a means of managing this independent structure (El Hedi 2015: 65).

5.2 Historical Representation

At the time of the Ottoman presence, the Algerians relied on traditional industries that represented a model of solidarity because the craft groups were considered as solidarity units, often integrated by inheritance.

This is why, in addition to grouping trades that allow families to subsist and make a profit, these groups offer the individual a cultural identity that will be attributed to them within the community to which they belong. In this regard, numerous studies indicate that the various transactions drawn up in court included the name of the individual's trade alongside his name, which became over time his official name. Thus, we say, for example, Ali Adh Dhbāgh (tanner) son of Mohamed Adh Dhbāgh (Hammach 2006: 38).

However, French colonization destroyed the conditions for reproducing the artisanal system and transformed artisans into peasants doing hard work. On the other hand, the colonists devoted the raw materials to European industries, which pushed the Algerians to abandon artisanal production.

Although this model of solidarity was destroyed by the occupier. The adoption of the socialist option by Algeria (1962–1989), encourage a set of mechanisms which are similar to solidarity economy, such as cooperatives, which are a socio-economic unit organized voluntarily by a group of people, on the basis of joint action, mutual support and collective responsibility for engaging in agricultural, industrial activity, commercial or tertiary. In accordance, with the principles of cooperation. The aim was to serve the economic and social interests of their members and of society in general (Al-Hayali 2017: 1), as well as mutual which is defined as a collective distribution of the costs of prevention against the risks to which member are exposed (Lavenseau and Smuerzinski 2006: 6). The adoption of these mechanisms was mainly aimed at ensuring social justice in Algeria in the socialist era.

5.3 Social Representation

Historical similarities can motivate the application of the solidarity economy, just as social similarities can stimulate mechanisms specific to a family economy that can give satisfactory results, especially since agriculture is still concentrated in areas, where tribal associations still play an important role and are generally marginalized areas. Therefore, following this model can create employment on the one hand and reduce regional disparity on the other.

The existence of numerous organizations set up to fight unemployment in Algeria such as the National Agency for Youth Employment, the National Agency for the Development of Investments, the Loan Guarantee Fund for Small and Medium-Sized Enterprises, the National Agency for the Management of Micro-Credit and other organizations can serve as a locomotive for the Algerian community economy. Likewise, the social assets of young entrepreneurs from craft families can represent a collective professional cultural identity for young entrepreneurs realizing economic profits on the one hand and reviving a cultural aspect destroyed by colonization. On the other hand, it is possible to exploit the "habous" as a venture capital for entrepreneurial incubator companies.

Additionally, social economy enterprises can provide mechanisms to reduce the size and impact of the transition to the neoliberal economy, particularly those linked to the informal economy. This can be done through social economy enterprises such as microfinance, micro insurance, subsidized health solidarity programs, social enterprises aimed at reintegrating disadvantaged groups of the population or undertaking work communities as well as a range of different types of cooperatives (El Hedi 2015: 76).

6 Conclusion

The historical dynamics of the Algerian economy before French colonization refers to two main factors: traditional/craft industries and agriculture. Colonization dismantled the relationship that linked the Algerian to his land, in particular its relation with his tribe and his craft, inherited from father to son and identified with it. It adopted an extractive economy model accompanied with destruction of the socio-economic structures of the Algerian economy dating from the pre-colonial period. The failure of Algerian development models, in the restoration of these socio-economic structures, contributed into the creation of the same disturbances, which characterized the period especially at the level of urbanization that continued on the same colonial model consecrating regional disparity.

The adoption of the neoliberal economy also created a distorted economic model that enshrined regional and social inequality, which, if continued, would lead to imbalances and social movements that could lead to the disruption of social, economic and even at the political level.

Therefore, the participation of society through the solidarity sector as another pillar of the economy would lead to the construction of the socio-economic foundations of the Algerian economy and could be a fundamental starting point for reducing the scale of the regional and social inequality.

However, this discourse remains theoretical if it does not lead to a convergence between public action and civil society by building a democratic society that constitutes the main channel for promoting solidarity and partnership. Thus, it is not useful to separate the two freedoms, political and economic, promoted by the neoliberal predominant discourse that lead to the continuity of economic dependence and the alliance between the oligarchic elites in the Third World countries and the elites industrial and financial in developed countries.

Above all, the flexibility of a solidarity economy would allow this model could be structured according to the specificities of Algeria, far from the ready-made solutions which are generally proposed by neoliberal institutions and which only lead to the reproduction of economic crises.

Bibliography

Akaba, A.: Management of Urban Policies in Algeria - Batna Case Study, University of Batna: Faculty of Law, Department of Political Science, Master thesis (2010). (in Arabic)

Al-Hayali, W.N.: Accounting of Cooperatives. Editions of the Arab Academy, Denmark (2017). (in Arabic)

Allard, J., Davidson, C.: Solidarity Economy: Building Alternatives for People and Planet. Change Maker Publication, Chicago (2008)

Allouche, Y.: Why Algeria's 'Black October' in 1988 defined its role in the Arab Spring, Middle East Monitor. https://www.middleeastmonitor.com/20161006-why-algerias-black-october-in-1988-defined-its-role-in-the-arab-spring/

Altayeb, A.: Urban transformations in mining areas in the arab maghreb, mining areas in Southern Tunisia as a Model. Insaniyat 42 (2008). (in Arabic

Algérie Presse Service: The average monthly salary is estimated at 39,000 dinars in 2016 (2017). (in Arabic). http://www.aps.dz/ar/economie/50626-39-000-2016

World Bank: Rural population (% of total population), World Bank database (2019). (in Arabic). https://data.albankaldawli.org/indicator/SP.RUR.TOTL.ZS?locations=DZ

Belamri, S.: Conviction of 612 mayors and 1,174 local voters in financial and real estate scandals. Echourouk 1820 (2006). (in Arabic)

Buehn, A., Karmann, A., Schneide, F.: Shadow economy and do-it-yourself activities: the german case. J. Inst. Theor. Econ. JITE (2009)

Boutelaji, I.: 14,000 demonstrations in Algeria each year!. Echourouk (2016). (in Arabic). https://bit.ly/2OU3ari

El Hedi, A.A.: Solidarity economy and social development: potential and reality in Mauritania, University of Tlemcen: Faculty of economic, commercial and management sciences, unpublished doctoral thesis (2015). (in Arabic)

Hammach, K.: The family in Algiers during the Ottoman era, University of Mentouri-Constantine: Faculty of Human and Social Sciences, Department of History, unpublished doctoral thesis (2006). (in Arabic)
Kahal, H.: The IMF recipe for Algeria as a calming agent for worsening economic crises, Al-Arabi 21 (2018). (in Arabic). https://bit.ly/2EaoG39
Kalthoum, M.: Urban development in Algiers from 1830 to 1939. Periodical Rev. Hist. Geogr. **3**(5) (2012). (in Arabic)
Labichi, E., Alkama, D.: Territorial organization and its impact on the spherical structure in Algeria, Barid al-Maarifa, University of Biskra **10** (2010). (in Arabic)
Lavenseau, D., Smuerzinski, E.: Social and solidarity economy in the urban community of Lille. Study report, National Institute of Statistics and Economic Studies (INSEE), France (2006)
Luca, R.: The Definition of Subsidy and State Aid: WTO and EC Law in Comparative Perspective. Oxford University Press, Oxford (2009)
Marwan, A.: Protests from the south in Algeria. Riot or legitimate claims? Al-Jazeera Net (2018). (in Arabic). https://bit.ly/2DbH5yp
Maurisse, M.: Powdered milk fraud ended in Switzerland. Swissinfo (2018).(in French). https://www.swissinfo.ch/fre/economie/crimes-%C3%A9conomiques_la-fraude-au-lait-en-poudre-finissait-en-suisse/44369212
Massaai, M.: The Policy of Economic Recovery in Algeria and its Impact on Growth. elbahith for academic studies (2014). (in Arabic)
Miri, Y.: Jarada: the city and society after the closure of the coal mine, Fes, Faculty of Letters and Human Sciences, Dhahr al-Mihraz, doctoral thesis in sociology (2004). (in Arabic)
Mokhtar, A.: Tax reforms and their role in achieving social justice in Algeria, University of Tlemcen, Faculty of Economics, unpublished doctoral thesis (2016). (in Arabic)
Mikaoui, M.: Tax expenditures in Algeria and the problem of tax evasion. Evaluative analytical study. Revue d'études fiscales **4**(2) (2015). (en arabe)
Oxford Business Group, the Report: Algeria: Oxford Business Group, Country Business Intelligence Reports. London, September 9, 2007 (2007)
Oxford Business Group, the Report: Algeria: Oxford Business Group, Country Business Intelligence Reports. London, October 10, 2010 (2010)
Smith, P.: Assessing the size of the underground economy: the Canadian statistical perspectives. Can. Econ. Observer, **11**(010) (1994)
Stora, B.: La guerre, l'histoire, la politique. Editions Michalon, Paris (1995)
Um Kalthoum, M.: French policy towards the Algerian Sahara, 1954–1962. University of Saida, Faculty of Social and Human Sciences, Department of Human Sciences, Master thesis (2013). (in Arabic)
World Health Organization: Total expenditure on health as % of GDP (2014). http://www.who.int/countries/dza/en/
World Population Review: Algeria Population 2018 (2019). http://worldpopulationreview.com/countries/algeria-population
Zouzou, R.: Rural migration in new social transformations in Algeria 1988–2008, University of Mentouri-Constantine: Faculty of Human and Social Sciences, Department of Sociology, unpublished doctoral thesis (2008). (in Arabic)

Social Technology as a Booster for the Social Innovation

Hamoul Tarik[1], Zair Wafia[2(✉)], and Kassoul Sofiane[1]

[1] University of Bechar, Bechar, Algeria
univbtarek@gmail.com, sofianekassoul@hotmail.fr
[2] University of Blida 2, Blida, Algeria
marketingblida@gmail.com

Abstract. The primary objective of this paper is to contribute to the existing literatures by comprehensively reviewing the definitions, concept, as well as the importance of social technology in the social innovations world. This review covers various definitions of technology and innovation, multi dimensions of social technology and how its acts to boost and generate a good social innovation. especially Within the contemporary climate of chaotic socio-economic change and institutional crisis, where increasing attention is being paid to experiences of social innovation Many forms of technology cycle models have been developed and utilized to identify new/convergent technologies and forecast social changes. This paper uses Gartner Group's Hype Cycle as an analysis tool to understand the impact and role of the social technology in the creation and the development of social innovation.

Keywords: Technologies · Innovation · Social technologies · Social innovation · Media technology

1 Introduction

The Social technology is a pun whose multiple meanings reflect the complexity of new understandings of 'the Social'. The many and diverse issues that connect 'technology' and 'Social' are indicated first by the increasing, and productive, convergences between science and technology studies on the one hand and sociology on the other. we observed that there has been a huge growth in the production of knowledge about economic life. sociologic and economic analysts produce assessments of current and future performance of firms, industrial sectors, and national and global economies by giving importances to the social sides and its roles. so, political activists, policy makers and social theorists have sought to develop accounts of the globalisation of sociologic activity, its influences and the weaknesses of policies and politics that confine themselves to the level of the nation state.

– **Aim of study:** The aim of this paper is to provide insight into the concept of social technologies, to develop its meaning in information and knowledge society by evaluating the impact of the social collaboration tools or technologies in social innovation.

K. Boussafi et al. (Eds.): MSENTS 2019, LNNS 162, pp. 76–90, 2021.
https://doi.org/10.1007/978-3-030-60933-7_5

so the Purpose is to present scholarly viewpoint on social technology as a research field and its impact on social innovation. more over, Social innovation requires extensive networking, communication, and collaboration among various social actors. This article presents an approach to fostering and supporting social innovation through the exploitation of multiple social technology.

We acknowledge the fact that the nature of technologies may change; therefore, our research focus is on enduring social practices and mechanisms by which transient technologies can support the innovations practices.

2 Social Technology

Technology, as it is defined in some books, means the processes by which an organization transforms labor, capital, materials, and information into products and services of greater value.[1] that why the Social technologies as a term, continue to change and grow popularity inside the society. Even though the term "social technology" is most commonly used to refer to new social media such as Twitter and Facebook, a redefinition of this concept based on the original definition is needed. Nowadays the concept of "social technology" has several aspects which destabilize the dominant image of technology. It emphasizes the social sciences and the humanities as society shapers, reconsiders the strength of "soft technologies" Based on the analysis of the scientific literature and empirical results in the Focus group as well as content analysis theoretical framework for defining social technologies was developed.[2] The concept of social technology generally is inseparable from the concept of information and communication technology. Thus this definition, despite of its popularity and wide application may be named as narrow approach of the general category of social technology. Some scientist suggest to disassociate from the informational technologies and reveal the meaning of social technology in wider scope, as all possible problems solving methods, when some negative social phenomena with a help of certain combination of tools and methods, is changed into more desirable in society. Thus, even this wider attitude (in distance with the compulsory connection with technological progress as a main characteristic of social technology), gives us the same keywords, for defining social technologies: innovative, more effective and changing the common processes.[3] This concept of technology therefore extends beyond engineering and manufacturing to encompass a range of marketing, investment, and managerial processes. Innovation refers to a change in one of these technologies.[4]

Social technology as a term popularity in academia grows fast, especially in relation to management, sociology, psychology, information technology and law disciplines.[5] In modern understanding of social technology, it could be applied for various purposes, such as decision making, knowledge sharing, etc. Social technologies can be defined as any technologies used for goals of sodium or with any social basis, including social hardware (traditional communication media), social software (computer mediated media), and

---

[1] See [1].
[2] See [2].
[3] See [3].
[4] See [1].
[5] See [4].

social media (social networking tools). social technologies "is a digital technologies used by people to interact socially and together to create, enhance, and exchange content". Social technologies distinguish themselves through the following three characteristics:

- they "are enabled by information technology";
- they "provide distributed rights to create, add, and/or modify content and communications";
- they "enable distributed access to consume content and communications".

Social technologies include a wide range of various technological instruments that can be used by people, private or public sector organizations, or as an interaction tool between them. They include many of the technologies that are classified as "social media", "Web 3.0", and "collaboration tools" (see Fig. 1).[6]

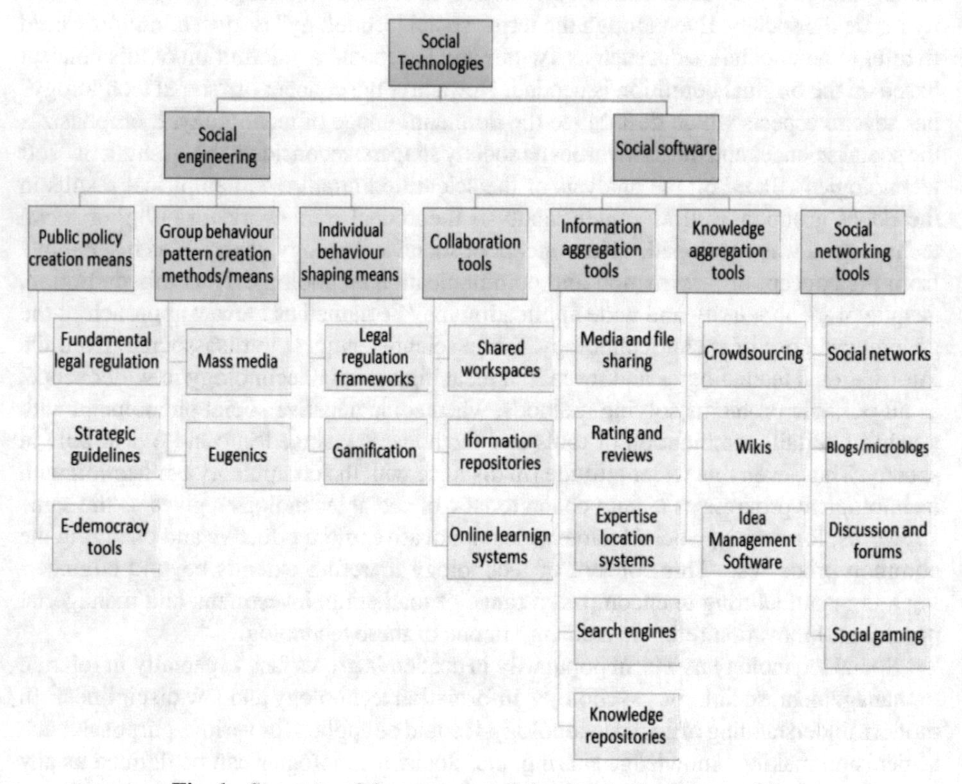

Fig. 1. Structure of the concept social technologies **Source:** [4].

[6] See [3].

3 Structure of the Concept "Social Technologies"

Social technologies when analysed often loses their true meaning due to endeavours to separate the 'social' and the 'technology'. Authors noted that

- "All technologies involve and shape the social" (Derksen et al., 2012)
- studies based on engineering perspective often use the term "sociotechnical" to denote the social aspects of technology and the deeply technological nature of society.
- Authors noted also that the Social technology concept referred to contemporary social media technologies which enable citizens, private interests, and others to form groups, develop coalitions, argue for change or influence, and then disband are powerful tools. they have been embraced by a variety of stakeholders, as well as by government agencies. Such technologies open the regulatory process to hundreds of thousands of citizens who never would have been able, or perhaps have even known, about the opportunity to offer their views before a new regulation is finalized.[7] Structure of the concept 'social technologies' observed in academic literature revealed complex structure encompassing variety of elements we can focus on some of them in Fig. 1 below.

As is showed above, the social technology term encompasses complex concept covering social engineering and social software elements under a hierarchical umbrella. This umbrella stands over such daily life elements as laws, mass media, social media platforms, etc. Social technology term is wide, old and covering all non-physical technologies.[8] Such separation is only conditional, because every tool, used in certain sphere, has a potentiality to be used in other one. For example: social networks, as a tool for communication, which may be defined as a set of socially relevant network members, connected with one or more relations, nowadays is widely used in marketing and in involvement of society into decision making processes.[9] Another researchers see that all listed social collaboration tools and technologies may be conditionally separated into three big groups: e-business tools, e-government tools and e-community tools as it is showed in Fig. 2 below. The majority of the mentioned social technologies have some common characteristic: the better accessibility and affordability, granted by the usage of collaboration tools.[10]

moreover, another limitation of this concept is its exclusive focus on the affordances of social technologies for knowledge practices in organizations.[11]

4 Innovation: Definitions and Concept

In general, the concept of "innovation" - a rather complex and multifaceted, his study of the subject of many studies, but, despite this, the generally accepted definition of

[7] See [4].
[8] See [4].
[9] See [3].
[10] See [3].
[11] See [5].

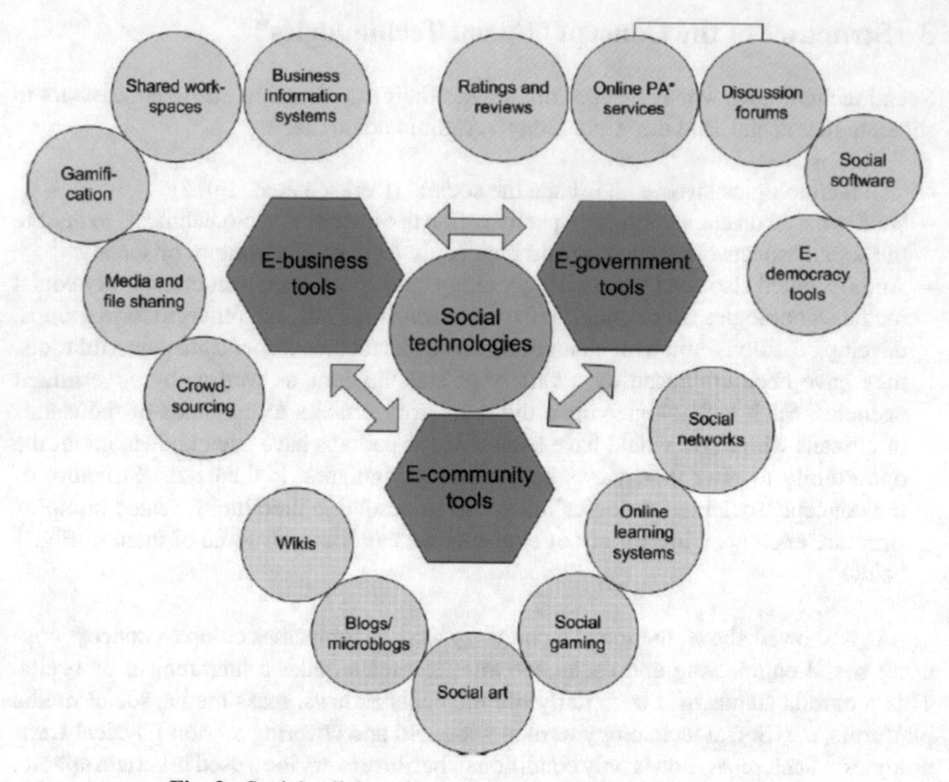

Fig. 2. Social collaboration tools and technologies **Source:** [2].

innovation in science does not exist. There are three main approaches to the consideration of the term. Schumpeter, which may be called the founder of the theory of innovation in the economy generally, regarded innovation as the economic impact of technological change, as the use of new combinations of existing productive forces to solve the problems of business.[12] Since Schumpeter, the concept of innovation has focused predominantly on economic and technical developments, whereas social sciences were particularly interested in the corresponding social processes and effects.[13] According to Twiss, innovation - a process that combines science, technology, economics and management, as it is to achieve novelty and extends from the emergence of the idea to its commercialization in the form of production, exchange, consumption. Afuah refers to innovation as new knowledge incorporated in products, processes, and services. He classifies innovations according to technological, market, and administrative or organizational characteristics.[14]

This may explain why social sciences, until this day, have been conducting empirical work on social innovations quite comprehensively, but without labelling them as such and, with few exceptions, without a concept of social innovation informed by social

[12] See [6].
[13] See [7].
[14] See [6].

theory. Technological innovations are elements of this continuous process and, due to the predominant patterns of imitation and invention, they have become the centre of attention.[15]

In general, There are three main approaches to the consideration of the term innovation. This classification is presented in the (Fig. 3) below.

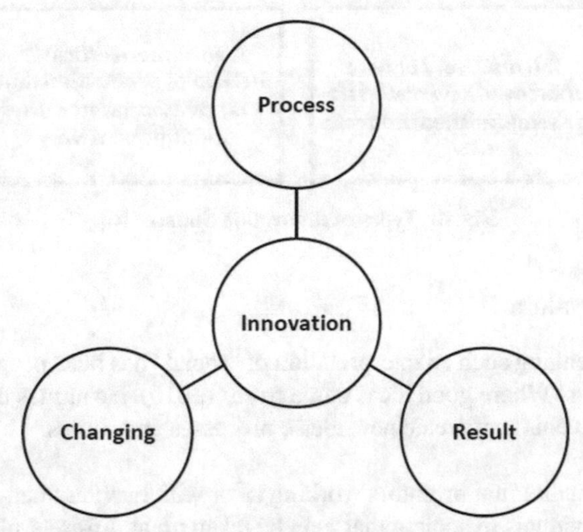

Fig. 3. Approaches to the definition of innovation **Source:** Siauliai A. (1979). The Essence Of The Concept Of "Innovation" As An Economic Category And Economic Systems Management, Electronic Scientific Journal. http://Www.Uecs.Ru, Date: 31.10.2013.

Innovation is divided into several dimensions but according to literatures, administrative and technological innovation are the most discussed in numerous of studies.[16]

4.1 Types and Classification of Innovation

Nowadays all economic processes are closely related to new technologies and innovations. The most important thing for scientists and developers to understand what kind of innovation should be used. for instance, Schumpeter identified the five types of innovation. The description of this model is presented on (Fig. 4) below.

15 See [7].
16 See [8].

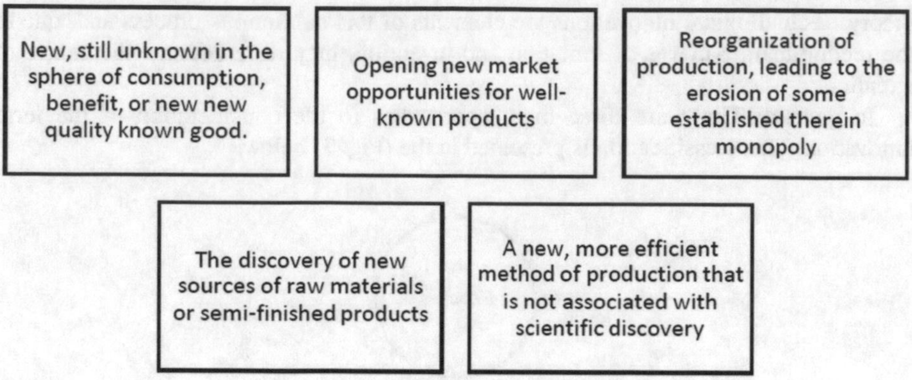

New, still unknown in the sphere of consumption, benefit, or new new quality known good.

Opening new market opportunities for well-known products

Reorganization of production, leading to the erosion of some established therein monopoly

The discovery of new sources of raw materials or semi-finished products

A new, more efficient method of production that is not associated with scientific discovery

Fig. 4. Types of innovation **Source:** [6].

4.2 Social Innovation

One way of challenging such an interpretation of "social" has been proposed by Johnson in his essay asking "Where good ideas come from" (2010). He argues that there are four different environments that create new ideas, processes and things:

(a) the ideas of individual inventors working as or with businessmen,
(b) ideas of individuals in society that may be taken up at different places,
(c) market-networked innovations, generated by (clusters of) enterprises and their R&D departments and finally,
(d) what he calls "the fourth quadrant" (2010, p. 213), non-market/networked movements inventions and actions making them practical innovations.[17]

Although as an academic concept, it is less wide-ranging, there still remains a broad range of interpretations. Some posit simply that it must constitute a new approach to a particular kind of problem. The Stanford Centre for Social Innovation, for example, describes it as "the process of inventing, securing support for, and implementing novel solutions to social needs and problems".[18] So, defined social innovation as both products and processes; ideas translated into practical approaches; new in the context where they appear. It is important sometime to use such a definition, rather than a more specific one, because one cannot clearly predict what comes out of even a very promising innovation in the course of its development. The problem with defining social innovation resides less in innovation" and much more in the meaning one attributes to "social". Studying the current literature on conceptualising and defining social innovations, one finds that "social" is mainly equated with improvement" (Phillis 2008), finding better answers to basic needs and more satisfying social relations (Moulaert 2010), and a range of other "good things".[19]

[17] See [9].
[18] See [9].
[19] See [9].

At least, The definition of social innovations is a bone of contention. In their overview written for the European Commission and the WILCO project, Jenson and Harrison have referred to social innovation as a "quasi-concept", a "hybrid, making use of empirical analysis and thereby benefitting from the legitimising aura of the scientific method, but simultaneously characterised by an indeterminate quality that makes it adaptable to a variety of situations and flexible enough to follow the twists and turns of policy, that everyday politics sometimes make necessary" (European Commission 2013, p. 16). Indeed, it has achieved the status of a buzzword in national and European policy circles. US President Obama established no less than two offices for social innovation. The EU has used the term to fund several initiatives. It is then little wonder that the meaning has diluted, sometimes referring to anything that is considered new and that is not technical.[20] (Fig 5)

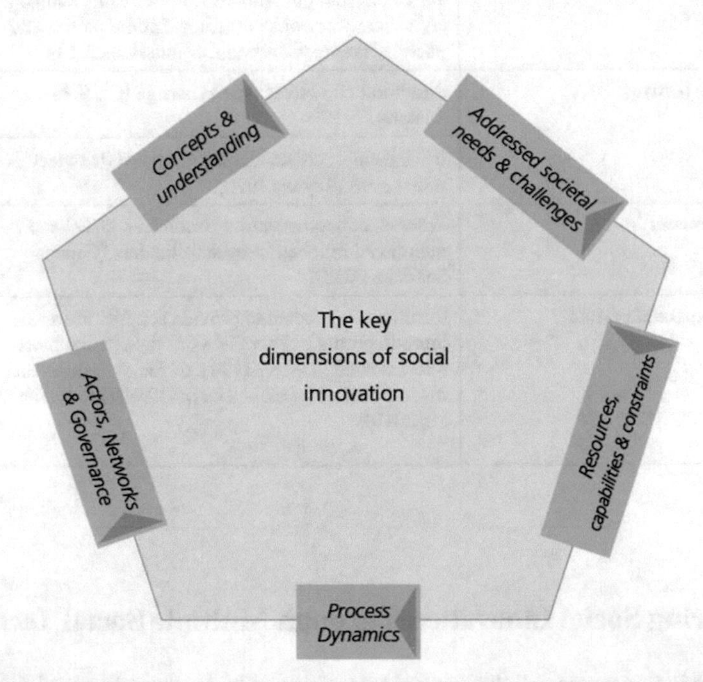

Fig. 5. The key dimensions of social innovation **Source:** Howaldt, Butzin, Domanski and Kaletka (2014).

[20] See [9].

5 The Key Dimensions and Characteristics of Social Innovation

– The Table 1 below summarize the social innovation characteristic:

Table 1. The social innovation characteristic: **Source:** Paulo Henrique De Souza Bermejo, Handbook of Research on Democratic Strategies and Citizen-Centered E-Government Services, IGI Global, USA, 2015, p:151.

	Social Innovation
Actors	Individuals (Lettice & Parekh, 2010), policymakers, foundations, entrepreneurs, philanthropists, social organizations (Murray et al., 2010), and governments (Pol & Ville, 2009); civil society organisations, local communities and puclic servants (Europen-Commission, 2013).
Objectives	Structural objectives: social change (Cajaiba-Santana, 2013).
	Instrumental: create technical articles that meet a social need (Taylor, 1970).
Process	Process: collective action (Neumeier, 2012) and intentional innovation by stakeholders (Cajaiba-Santana, 2013).
Expected results	Results are expected to provide benefits to society through products, processes or services that meet a social need (Taylor, 1970), or social changes that institutionalize a new social practice (Howaldt et al., 2010).

6 Fostering Social Innovation Through Multiple Social Technology

This section in this paper will discuss the role of social technology in a social innovation for a route to success. This will be presented and discussed in the following points:

- The Social technology is intended to help a wide range of managers, consultants, and academics in many businesses, high tech or low, in slowly evolving or rapidly changing environments and allow to understand many of life's most challenging endeavors, own a great value in coming to grips with "the way to world works," and in managing innovative efforts in ways that accommodate such forces.
- Social technology allow to understand the trajectories of Market Need versus Technology Improvement and the elimination of the observation that technologies can progress faster than market demand, illustrated in Fig. 1, means that in their efforts to provide better products than their competitors and earn higher prices and margins,

suppliers often "overshoot" their market: They give customers more than they need or ultimately are willing to pay for. And more importantly, it means that disruptive technologies that may underperform today, relative to what users in the market demand, may be fully performance competitive in that same market tomorrow.

– The concept of innovation was initially focused on the private sector, and consisted in new combinations of production factors (according to the Schumpeterian definitions [Schumpeter, 1967]), leading to new products and services, or/and new production processes, and having mainly economic objectives and rationale (aiming to increase the sales revenues and profits of innovating firms). However, some fundamental changes in the economy and the society that took place in the first decade of the 21st century lead to serious and complex social problems affecting large citizens' groups, which could not be addressed by existing market offerings or government services, and necessitated a new form of innovation, referred to as "social innovation," which has social objectives and rationales (rather than economic ones), and is based on cooperation of multiple social actors.[21] (Fig. 6)

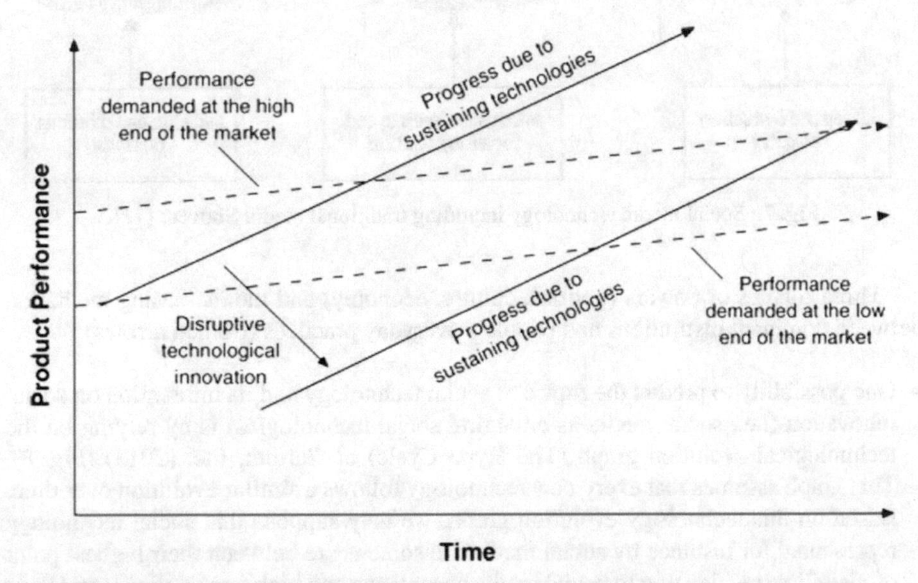

Fig. 6. The impact of sustaining and disruptive technological change **Source:** [10].

There is a general effect of new technology on the media subsystem in all three orders or powers, and a specific one that pushes new technology into the other two orders but also into the cultural sphere itself, adding to how cultural content (including scientific communication that could be developed to an innovations) is mediated, which includes the growing place of scientific knowledge in society and the increasing everyday uses of media technology in everyday cultural life as part of an overall cultural development that would help to create a social innovation. This section (and especially the preceding

[21] See [10].

paragraph) made a complex argument; again, it is crucial to what follows, and also to the overall aim of offering a comprehensive theory of the role of the internet which can be considered as one of the most social technology (social media technology including traditional media) used in society. Figure 7 shows how the role of social technology, as media technology, is both part of culture but also – as techno-science – drives change within culture and the other spheres, via dashed arrows: technology (or techno science) is a separate force.[22]

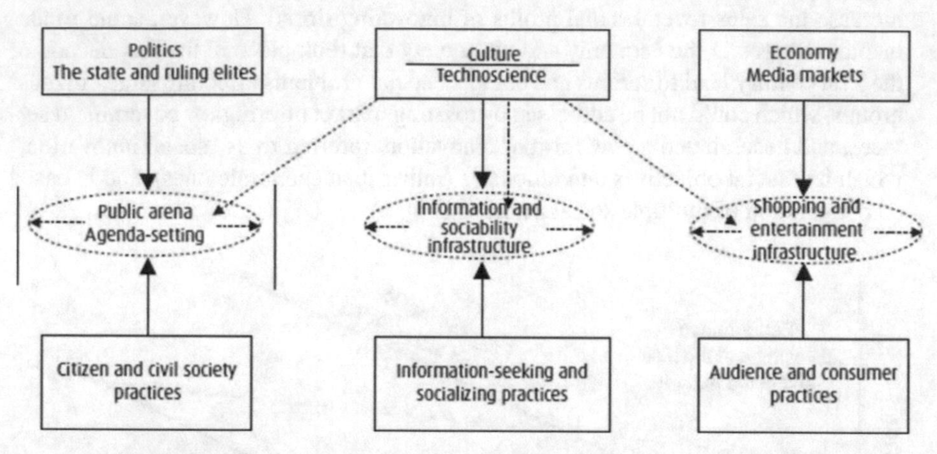

Fig. 7. Social media technology including traditional media **Source:** [11].

Three spheres or powers (politics, culture, economy) and the increasing mediation between dominant institutions and people's everyday practices (dashed arrows).

- One possibility to predict the future of social technology and its influention on social innovation (i.e., social media as emerging social technologies) is by relying on the technological evolution graph (The Hype Cycle) of Gartner, Inc. (2013) (Fig. 8). This graph assumes that every new technology follows a similar evolution over time. Based on this technology evolution graph, we may suppose that social technology represented for instance by social media are somewhere between their highest point of visibility (i.e., leading to social media acceptance and high expectations) and their lowest point of visibility (i.e., with organizations getting disillusioned after losing money in projects that use social media in an inappropriate way). For instance, the technology evolution graph can explain both the current social media hype and the social media bloopers of today. Either organizations will find a more mature way of working with social media or the social technologies will disappear (or will only stay for a while with a low productivity level). Only in the first situation, organizations will better understand how to properly use social media, resulting in a relatively steady visibility level and with a relatively high productivity.[23]

[22] See [11].
[23] See [12].

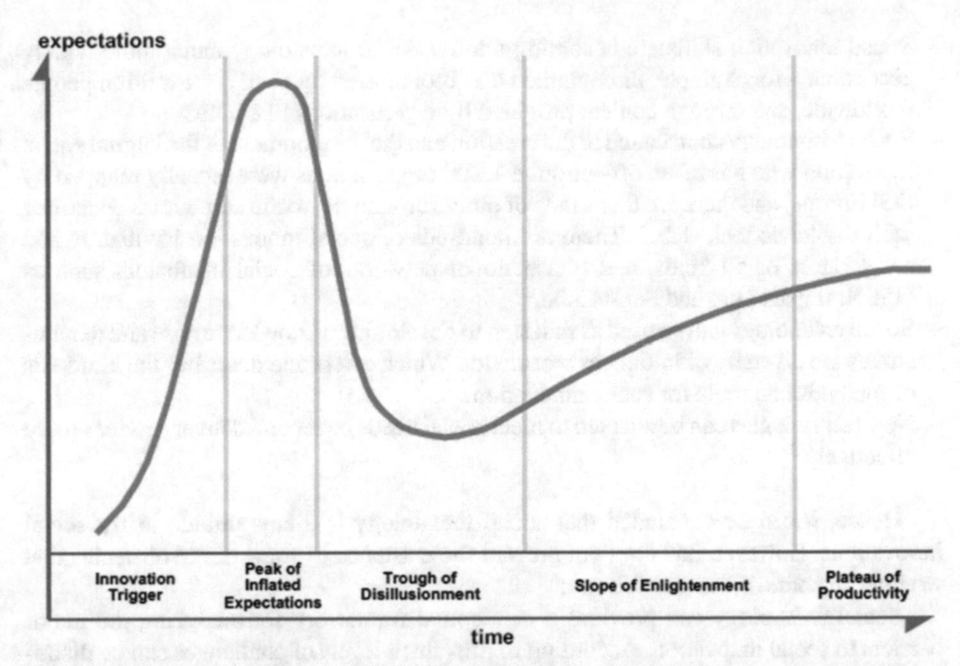

Fig. 8. The Hype Cycle **Source**: Jackie fenn and others, understanding gartner's hype cycles, gartner corporate headquarters foundational document, USA, 2017, p:04.

Academically, there are also various reasons to be careful with the term. To begin with, the concept of social innovation is poorly defined and demarcated. Furthermore, innovations are too often presented as "pearls without an oyster": They are pretty to look at, but we do not know where they come from. so the social technology can guide us to the sources of innovations.[24] add to that, the social technologies facilitate informal knowledge sharing in the workplace.

– social technologies (Social media for instance) provide advantages for infusion of innovative ideas. The uses of social media contribute to the transfer of innovative idea through scaffolding horizon-broadening practices, primarily by helping people grow the number of weak ties and also leverage this network to get one's ideas circulating. many instances, social media are useful for nurturing weak ties, and therefore they facilitate the transfer of innovative knowledge. While people draw on strong ties for work-related advice that directly influences their work in reaching out practices, they also benefit from creative and innovative ideas shared through weak ties enabled by social technologies like the social media.[25]
– The social anthropology might be very helpful if we want to understand better the links between Social Innovation and Social technology.

[24] See [9].
[25] See [5].

- Social innovation skills are becoming via the Social technology much more widely accessible – for example, through the DIY Toolkit used by nearly one million people worldwide, and through content provided by organisations like IDEO.
- Social technology contributed in the creation and the development of the Digital social innovation who has taken off – around 1,200 organisations were recently mapped by DSI Europe, and there are thousands of others around the world sometimes described with the 'civic tech' label. There are hundreds of social innovation incubators and accelerators of all kinds, and transnational networks of social incubators such as GSEN, Impact Hub and SenseCube.[26]
- Social technology can be used as first step to developing a knowledge base and demonstrates the diversity of initiatives worldwide Which offers and describes the hundreds of methods and tools for social innovation.
- New technologies can be adapted to meet social needs better and deliver services more effectively.[27]

Hence, it can be concluded that social technology is a key attitude in for social innovations. However, this does not prevent the existence of some of the obstacles that we refer to some of them as follows:

Social technology can provides a conceptual framework for analyzing the major barriers to social innovation. According to this, three levels of challenges can be distinguished: environmental challenges (lack of finance, non-financial resources, networks, etc.); underlying challenges (difficulties of adopting an open innovation paradigm; measurement-related challenges) and actor-related challenges (lack of capacities and competencies).

Technology-based innovation environmental analysis (social technology) had indicate the need to focus on the reinforcement and replenishment of factors that encourage social innovation, and the factors that weaken innovation and remove the factors that impede social innovation. so, the development and growth of social innovation is impeded by factors such as limited access to finances, poorly developed networks and intermediaries and limited skills and support structures.[28]

Finally, Social innovation cannot be naively conceived as a cheap do-it-yourself alternative to universalistic welfare provision and protection. However, under specific conditions that still need to be carefully analysed and theorized, social innovation may become a powerful tool for confronting the disintegrating developments of our times and used for Fostering Social Innovation.

7 Conclusion

Ubiquitous technology has changed the way people work, live and play. In contemporary, societies use information and communication technology (ICT) to search for information, make purchases, apply for jobs, share opinions, and stay in touch with friends and relatives. At the same time, contemporary this social technology can be considered as a

[26] See [13].
[27] See [14].
[28] See [15].

boon for the social innovations, its promotion and its development. however the importance and role of social tech still need to be carefully analyzed and theorized. moreover, social innovation may become a powerful tool for confronting the disintegrating developments of our times. The application of technologies represents a heavily debated aspects of societies. we had always the idea that "the Technology is there to Support Humanity and support innovation," that why the social innovation, as mentioned above, had been created via an extensive networking, communication, and collaboration among various social technology actors. These critical preconditions of social innovation are strongly associated with the fundamentals characteristics of the recently emerged Web 2.0 social media: Online community building and social networking, user generated social multimedia content intended to be shared with other users, rated and commented by them, and extensive users' interaction and collaboration.

In light of the increasing importance of social innovation, this paper tried to focuses on a theoretically sound concept of social technology as a precondition for the development of an integrated theory of socio-technological innovation in which social innovation is more than a mere appendage, side effect and result of technological innovation. Only by taking into account the unique properties and specifics of social technology, it will be possible to understand the systemic connection and interdependence of social technology and social innovation processes and analyse the relationship between social innovation and social change. more than that and against the background of the emergence of a new innovation paradigm, it becomes more important to devote greater attention to social technology as a booster for social innovation and as mechanism of change.

References

1. Christensen, C.M.: The Innovator's Dilemma: When New Technologies Cause Great Firms to Fail, p. 11. Harvard Business School Publishing, Boston (2000)
2. Skarzauskiene, A., et al.: Defining social technologies. In: 4th international conference on Information systems management and evaluation, p. 239. RMIT University Vietnam, Ho Chi Ming City, January 2013
3. Skarzauskiene, A., et al.: Defining social technologies: evaluation of social collaboration tools and technologies. Electron. J. Inf. Syst. Eval. **16**(3), 235 (2013)
4. Tamošiūnaitė, R.: What approach is suitable for social technology research? Contemp. Res. Organ. Manage. Adm. **3**(1), 97–98 (2015)
5. Jarrah, M.H., Sawyer, S.: Social technologies, informal knowledge practices, and the enterprise. J. Organ. Comput. Electron. Commer. **23**(1–2), 34–98 (2013). University of Berlin, Germany
6. Kogabayev, T., Maziliauskas, A.: Social technologies, informal knowledge practices, and the enterprise. Holistica **8**(1), 60–98 (2017). Aleksandras Stulginskis University, Lithuania
7. Howaldt, J., et al.: Social innovation: towards a new innovation paradigm, (mackenzie management review, universidade presbiteriana mackenzie), p. 22 (2016)
8. M.Z, Nur Fadiah., et al.: Defining the concept of innovation and firm innovativeness: a critical analysis from resource-based view perspective. Int. J. Bus. Manage. **11**(3), 91–98 (2016). Canadian Center of Science and Education
9. Brandsen, T., et al.: Social Innovations in the Urban Context, p:05. Springer New York (2016)
10. Charalabidis, Y.: Fostering social innovation through multiple social media combinations. Int. J. Bus. Manage. **31**(3), 225–239 (2014). Taylor & Francis Group, LLC

11. Schroeder, R.: Social Theory after the Internet Media, Technology and Globalization, p. 12. UCL Press, University College, London (2018)
12. Looy, A.V.: Social Media Management Technologies and Strategies for Creating Business Value. Springer Texts in Business and Economics, p. 09. Springer, London (2016)
13. Mulgan, G.: Social Innovation – the Last and Next Decade, p:02. Nesta, London (2017)
14. Murray, R., et al.: The open book of social innovation, social innovator series: ways to design, develop and grow social innovation, the young foundation, p. 15 (2010)
15. Bilevičienė, T., et al.: Innovative trends in human resources management. Econ. Sociol. **8**(4), 107 (2015). Recent issues in economic development

University-Enterprise Cooperation: Determinants and Impacts

Dorra Mahfoudh[1](✉), Younes Boujelbene[1], and Jean-Pierre Mathieu[2]

[1] Faculty of Economics and Management of Sfax, Sfax, Tunisia
dorram17@yahoo.fr, younes.boujelbene@gmail.com
[2] Management Sciences, Paris, France
jpmnant@gmail.com

Abstract. This research presents a state of the art in university-business cooperation. First, it presents the factors that determine the propensity of the business sector and the university to deepen a cooperative relationship in the promotion of research and innovation. The quest to improve competitiveness is a determining factor in the success of this cooperation with the university, conditioned by a strong absorption capacity. On the university side, our results show that strategic orientation and the promotion of technology transfer are important factors that encourage collaboration with the company. Finally, relational collaboration mechanisms have a positive and significant impact on business innovation.

Keywords: Knowledge transfer · University-business collaboration · Innovation

1 Introduction

In a knowledge-based economy, science is increasingly influencing innovation. Thus, the extent and intensity of business-university relationships is seen as a major factor contributing to the promotion of innovation, whether at the level of enterprise or at the level of countries (OECD 2002). The OECD, in its recent innovation strategy (OECD 2010) recognizes that innovation is a very interactive process of collaboration within a network bringing together an increasing number of actors, institutions and users. In the same context, the Triple Helix model, popularized by Etzkowitz and Leydessdorff (1997) emphasizes the increased interaction between the different institutional actors in the innovation systems of industrial economies, especially the universities, the industry and the government.

A big interest was then put on the collaboration between the university, the company and the State.

The universities hold the key to accessing the knowledge economy thanks to their position at the intersection of research, education and innovation. (Shane 2005). Universities then assume a leading role in the creation of technological innovation and are considered to be engines of economic growth (Etzkowitz et al. 2000; Etzkowitz and

K. Boussafi et al. (Eds.): MSENTS 2019, LNNS 162, pp. 91–121, 2021.
https://doi.org/10.1007/978-3-030-60933-7_6

Leydesdorff 2000; Audretsch et al. 2013). The universities role has then evolved during the recent years. Indeed, According to Gunasekara (2004), universities were once described as "ivory tower" institutions focusing only on traditional academic practises in teaching and research, with little serious commitment to answer questions in relation to the socio-economic environment in which they operate. However, the role of universities has become more important with the emergence of the knowledge economy. Today, universities are gradually seen as the powerful engines of innovation and change in science and technology and in other creative disciplines (Sharma et al. 2006). In the development of the knowledge-based economy, universities should play an important role as facilitators, or even leaders of economic and social development as well as regional innovation systems (Gunasekara 2004). It is therefore important that policies promoting regional innovation processes include capacity-reinforcement programs for universities and emphasize the intensity, quality and socio-economic relevance of the researches carried out at the universities (Fritsch and Slavtchev 2007). The driving forces behind the emergence of this new role are the collaborative networks that link the private industrial enterprises, the entrepreneurial universities, the governmental organizations and the other public bodies (Etzkowitz and Kloftten 2005).

As part of this research, we have chosen to focus our interest on the issue of the relationship between the enterprise and the university with the state assistance and how to strengthen and develop innovation.

These are the unifying questions of our research, which needs a qualitative and quantitative exploration of the complexity of the University-Enterprise partnership phenomenon and the state assistance.

We therefore propose to answer,in the context of this work, the following research questions:

- What are the factors linked to the company, the university and the State which can encourage collaboration?
- What are the factors that can motivate both the industrialist and the academic to collaborate?
- What is the impact of this collaboration on business innovation in Tunisia?

2 State of the Art

2.1 Knowledge Economy

The term knowledge economy was born out of an awareness of the role of knowledge and technology in the economic growth.

Knowledge, as a human capital and included in technologies, has always been at the centre of economic development.

The knowledge economy is defined by the OECD as a system in which the production, use and distribution of information and knowledge are essential to the process of economic growth (OECD 1996).

2.2 Model of the Triple Helix

Among the theoretical models of innovation containing a knowledge-based economy (KBE), we quote the triple helix model developed by Etzkowitz and Leydessdorff (1997)

A triple helix regime usually begins when the university, industry and government enter into a reciprocal relationship in which each tries to improve the performance of the other.

Each institutional sphere is therefore more likely to become a creative source of innovation and to foster the emergence of creativity that manifests itself in other spirals (Figs. 1 and 2).

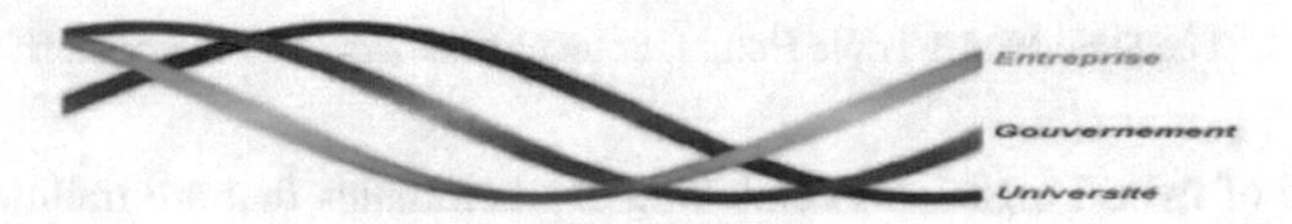

Fig. 1. Model of the Triple Helix

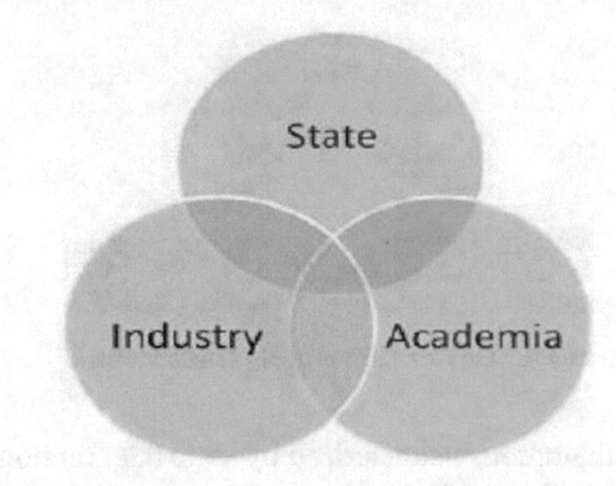

Fig. 2. Triple Heilix model Source: Etzkowitz (2002, 2004), Etzkowitz and Leydesdorff (1998, 2000)

Different forms of triple helix have been proposed by Etzkowitz and Leydesdorff (2000) (Fig. 3).

The government plans, controls, and directs the relationship between industry and academia in pursuit of innovation.

The industry is considered the champion of innovation.

While the role of the university is mainly reduced to teaching and research

Under this model, the potential for exploiting the knowledge generated by universities is limited as university teaching and research tends to be far from the needs of industry and even the universities themselves have little or no incentive to engage in the commercialization of their research (Bercovitz and Feldman 2006; Goldfarb 2003; Leydesdorff 2013). This unprecedented configuration, characterized by the meeting of the three major traditional poles in its centre, would offer the opportunity to see the

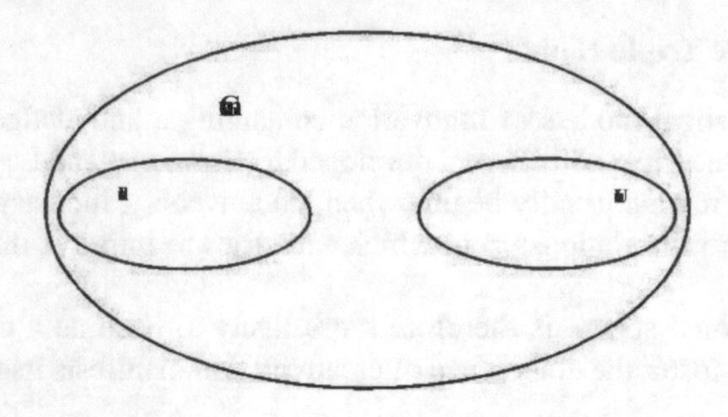

Source: Etzkowitz and Leydesdorff, 2000

Fig. 3. The State Model Triple Helix I. Source: Etzkowitz and Leydesdorff, 2000

appearance of hybrid organisms combining characteristics that are traditionally associated with one or other of the three poles concerned and thus making it possible to optimize the innovation objective (Etzkowitz and Leydesdorff 2000) (Fig. 4).

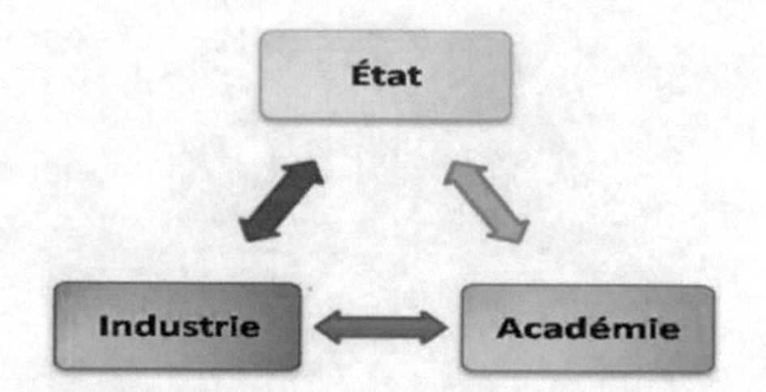

Fig. 4. Laissez faire model, Triple Helix II

A laissez-faire configuration, characterized by state intervention, is limited because the industry considers itself as a driver of innovation.

The other two spheres play the role of auxiliary structures and have a limited role in innovation.

The University acts mainly as a provider of qualified human capital and the government acts mainly as a regulator of social and economic mechanisms. (for example, the United States and some Western European countries). The absence of a synergistic relationship between institutional spheres means that the role of government in harnessing innovation is limited to addressing market failures, while universities engage in basic research and training of the workforce (Zheng and Harris 2007). Even companies integrated in the same industry operate independently of each other and are linked only through the market. Here too, the industry is considered to be the driving force for innovation with the two other institutional spheres as auxiliary structures (Etzkowitz 2003) (Fig. 5).

This hybrid model emphasizes the construction of overlap and relatively interdependent relationships between the three spheres.

Each institutional sphere maintains its own characteristics while assuming the role of others.

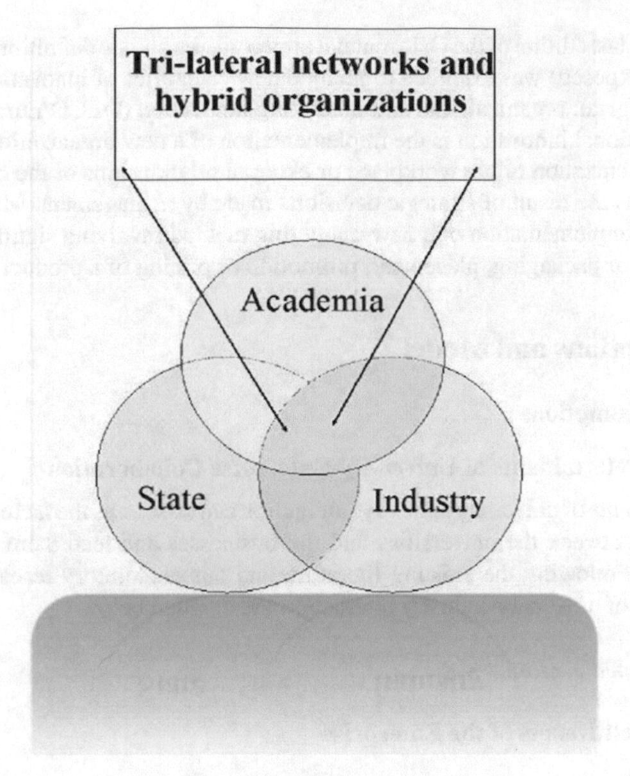

Fig. 5. The Hybrid Model: Triple Helix III

This configuration, characterized by the meeting at its centre of the three major poles, would offer the opportunity to see the appearance of hybrid organisms combining characteristics traditionally associated with each pole and thus making it possible to optimize the objective of innovation.

The triple helix hybrid model represents a combination of statist and laissez-faire models.

The hybrid model is a network that encourages the movement towards relations of mutual collaboration and links between the three main institutional spheres and other organizations in which the innovation policy is the result of their interactions.

2.3 Innovation

Innovation is then considered to be an interactive process (Lundvall and Nielsen 1999), and is the result of complex interactions between various actors and institutions (OECD 1996).

Innovation results in a system of internal interactions between different functions and distinct actors, in which experience and knowledge are mutually reinforced and accumulated.

When the last edition of the Oslo manual appeared, the classic definition of innovation (previously exposed) was extended to include new categories of innovations, known as non-technological: organizational and marketing innovation (OECD/Eurostat 2005).

Organizational innovation is the implementation of a new organizational method in practices, organization of the workplace or external relationships of the company. This new method is the result of strategic decisions made by management. Marketing innovation is the implementation of a new marketing method involving significant changes in the design or packaging, placement, promotion or pricing of a product.

3 Assumptions and Model

3.1 The Assumptions

3.1.1 The Determinants of University-Enterprise Collaboration

The determinants of university-industry interaction can be seen as the factors that shorten the distance between the universities and the businesses and lead them to establish a relationship. Following the existing literature and our exploratory research, the main determinants of university-industry interaction are detailed below

Business Related Factors

- **The competitiveness of the Enterprise**

Collaboration is the act of working with others to accomplish tasks and achieve common goals. It is a recursive process, in which two or more people or organizations work together. In particular, collaborating companies can gain more resources, recognition and rewards facing the competition for limited resources. Liao et al. (2017).

We can see that a company cannot become competitive overnight, and that to become competitive, the business must know how to make the best use of the internal and external factors available to it and to chart its course in the right direction. Companies are therefore constantly facing competition from one another. They, therefore,seek to improve their ability to remedy to this competition, that is to say: to improve their competitiveness.

According to Henri Spitzki (1995): “a company is competitive when it is able to sustain itself and voluntarily on a competitive and evolving market, by achieving a profit rate that is at least equal to the rate required by the financing of these Goals “.

The company’s competitiveness is therefore the ability to provide products and services more effectively and efficiently than the competent competitors. The company’s competitiveness is based on four developments:

- Technology development.
- Product development.
- Marketing development.
- Human resources development.

These developments can be enhanced by the expertise and knowledge of the company. The company cooperates with other expert companies, cooperates with universities for expertise, cooperates with other organizations, buys technologies from other companies or universities and gets advice from academics or experts.

We then propose the following hypothesis:

H1: *Companies looking to strengthen their competitiveness are more favourable to collaboration*

V' **The company's absorption capacity:**

Absorption capacity is defined as "the capacity to acquire, assimilate, transform and exploit knowledge" (Cohen and Levinthal, 1990).

The concept of absorption capacity focuses on the importance of a stock of prior knowledge to effectively absorb the bad consequences by cooperating, and stresses that internal technological capacities are necessary to benefit optimally from R&D cooperation. Bellucci1 and Pinocchio (2015).

Companies can effectively benefit from the technological flows resulting from collaborative research only with the capacities that are necessary to identify, assimilate, and develop useful external knowledge, as a result, some empirical studies have provided evidence that absorption capacity facilitates the transfer of knowledge between organizations (Lane et al. 2001). Although the absorption capacity is applied to all forms of cooperation, it is particularly important for the interactions with universities and other research institutions.

Lin et al. (2002) have also shown that companies cannot successfully assimilate and apply external knowledge without a very large absorption capacity. They explore the essential factors of absorption capacity through its impact on the transfer performance (of technology for example), and note significant associations between the absorption capacity and factors such as the distribution channels of technology, interaction mechanisms, and R&D resources.

As also underlined by (Bishop et al. 2011) and demonstrated by Garcia-Perez-De-Lema et al. (2016), the collaborative strategies of SMEs depend on two main factors: the absorption capacity of companies and their ability to develop personal relationships within their environment.

Therefore, we have further drawn the following hypothesis:

H2: *The absorption capacity of a company is a determining factor in collaboration with the university.*

Factors Related to the University V' **The strategic orientation of the University**

Some universities have better policies and structures in place to support and encourage collaborative research. Muriithia et al. (2017) mentioned that these differences were noted in the policies related to research training, the visibility of researchers, the accessibility of research information and questions concerning the ethics of research.

Belkhoja and Landry (2007) also mentioned that strategic positioning in a market is linked to the success of past collaboration experiences. The success of the research activities carried out by the researcher appears as an indicator of their strategic positioning with regard to the researcher's collaborative partners and the competition. The more past experiences which involve the use of research results have contributed to the development of new innovations, the more the researcher will be able to collaborate again. For this purpose, we can then assume in our research that researchers, who have adopted research projects for three years focusing on the needs of users such as private companies and government agencies and who have already participated in the creation of new products, have more chance to collaborate with the company.

We then propose the following hypothesis:

H3: *Strategic factors positively influence collaboration with the university.*

- **Promotion of technology transfer:**

Interaction between universities and businesses results from the need of the productive sector to develop a new technology, a new product or process, or even when there is an invention mature enough to be transferred from university to society, which is one of the possible interactions (Sankat et al. 2007).

According to C. Chais et al (2017), the main aspects related to technology transfer are: the need to professionalize and train ICT, the need to protect the intellectual property generated in universities, develop an entrepreneurial university to promote the culture of innovation, create internal innovation policies and map transfer processes to reduce bureaucracy in these activities.

Regarding cooperation between agents, Gorschek et al. (2006) proposed a university-industry technology transfer model based on a research calendar composed of several phases: the identification of potential areas for improvement according to the demands of the company through an observation s and evaluation process; formulate problems to be solved by studying the theoretical framework; propose solutions in collaboration with the company; develop laboratory validation; perform dynamic validation at a semi-industrial level; and gradually obtaining solutions, thus leaving the door open to other changes and proposals. According to these authors, the job of researchers is not just to do research, but to transfer technology.

Al-Agtash and Al-Fahoum (2008) have developed a model that represents a structure that aims to bridge the gap between academia and industry. The model offers a rich internal collaborative environment which brings together technical and commercial faculties in the pursuit of projects with the aim of cultivating collaboration with industrial and commercial partners, updating the knowledge base of the university and aligning the skills and knowledge of students with real and immediate industry needs.

The importance of organizational aspects in technology transfer has also been emphasized by authors such as Siegel et al. (Siegel et al. 2003a, b), who drew attention to the practices of university technology transfer offices acting as intermediaries and to the cultural barriers between universities and industry. The problems associated with technology transfer offices and their different organization (internal, external or mixed) are of great importance for the impact on the results of cooperation which implies the need to propose measures of their efficiency (Adams et al., 2001; Anderson 2007). This is particularly important due to the fact that universities are responsible for much of the technology-driven models (for example, University of Texas and Virginia Tech's Business Research Centre) and scientific parks (Research Triangle Park, Silicon Valley, etc.). At the European level, and in particular in the United Kingdom, the Technology Transfer offices are now considered to be the preferred approach for interaction between university and industry.

While many variations may be encountered, a technology transfer office should generally be staffed by professional knowledge transfer experts, to develop and execute the research institution's strategy for working with industry and users of research results, to help to identify, assess and (where appropriate) protect intellectual property, to advise on commercial matters, promote the use of inventions and other research and development results and disseminate information. SumChau et al. (2016).

Based on this evidence, the hypothesis to test is as follows:

H4: *Universities that aim to promote technology transfer are more favourable to collaboration.*

Factors Linked to State Incentives

- **State support**

The government R&D policy is also highlighted as crucial for the transfer of technology from public research organizations to industry. The Bayh-Dole Law of the United States in the 1980s is often mentioned as the legislation to facilitate the growth of university patents; the law stipulated that the results of university research carried out by public funds belong to universities. Since then, a number of OECD countries have imitated the adoption of similar laws in order to use university research for commercial benefits (OECD 2003). Government R&D support is useful for businesses that need external partnerships, but have financial or networking problems. The government can provide these companies with capital enabling them to acquire basic university technologies or the possibility of collaborating on research projects (Mohnen and Hoareau 2003). Mohnen and Hoareau (2003) and Capron and Cinccera (2003) show that companies, which use government support measures, tend to cooperate with these public research organizations.

Therefore, the role played by the government in the formulation of public policies is thus crucial for the interaction between the university and industry. By creating laws and setting standards, the government facilitates interaction between universities and businesses, encourages innovation and protects property rights (Mansfield, 1996). The triple helix model proposed by Etzkowitz (2003) presents three means of government influence on university-industry interaction: controller, regulator and financial support.

We then propose the following hypotheses:

H5: *State support facilitates the propensity to collaborate with the company*
H6: *State support facilitates the propensity to collaborate at the university.*

Factors Linked to the Motivations of the Researcher and the Industrialist The question of motivation is one of the major challenges of any collaboration. Two actors of the considered relationships have a completely different nature, so they have different research and collaboration objectives in general.

In addition, since the motivations for researchers who engage in collaboration are different from those of the industrialist in certain respects, we then propose in a first phase the hypotheses relating to the motivations of the industrialist and in a second phase the hypotheses relating to the motivations of the researchers.

1) Factors related to the motivations of the industrialist

The enterprises theory based on resources indicates that the internal resources play an important role in a firm's propensity for growth (Penrose, 1959, Richardson, 1972). If internal resources are constraining, it makes sense for a company to engage in cooperation with external partners to gain access to additional resources, such as capital, technology and human capital. Industry-university cooperation has been discussed as a type of R & D cooperation in these contexts. According to Geisler (1995), the more they are recognised as interdependent in terms of resources, the more the university and the company can establish partnerships. We then propose the following hypothesis:

H7: The engagement of companies in interactions with the university is motivated by different reasons.

Through our review of the literature and our exploratory analysis, we identified the most common reasons for business motivation and we then proposed the underlying hypotheses:

Access to research and the discovery of new knowledge and opportunities are considered as general motivations for companies (Bonaccorsi and Piccaluga 1994, Santoro and Gopalakrishnan 2000).

H7a: *Access to research is considered to be one of the motivational reasons that most encourage companies to interact with universities*

Problem solving (Bonaccorsi and Piccaluga 1994, Rappert et al. 2000) or assistance with general and specific problems is another motivation for companies to interact with universities.

H7b: *Solving technical problems is considered to be one of the motivational reasons that most encourage companies to interact with universities*

The outsourcing of R & D activities, reduction of risk, sharing of cost and access to public research money are motivations which are also mentioned by (Bonaccorsi and Piccaluga 1994). Some authors mention a tendency to outsourcing in R & D, in particular in technologies or in specific phases of technological development, such as prototype testing. With regard to cost sharing and access to public research money, some have stressed this in particular for small businesses, which have limited capacities and resources for internal R & D. In empirical analyzes, Belderbos et al. (Belderbos et al. 2004a, b) and Veugelers et al. (2005) reveal that the cost-sharing objective is one of the main factors influencing the company's decision to cooperate with the university. Companies collaborate with others for a variety of reasons, and cost or risk sharing is one of these reasons. This makes sense because R & D projects often involve high costs and risks and uncertainty (Jensen et al. 2003). Narula (2004) also concluded that due to the high costs and risks associated with setting up and managing a collaboration or alliance, SMEs prefer to outsource or externalise R&D activities in cooperation or collaboration.

H7c. *The outsourcing of Research and Development activities is one of the motivational reasons that most drives the company to collaborate with the university*
H7d. *Cost sharing is considered to be one of the motivational reasons that most motivate companies to interact with universities*

2) Factors related to the motivations of the researcher

Working with industry is a discretionary behaviour for academics. Research on the individual determinants of academic behaviour regarding the interaction with industry has given rise to conflicting opinions on the relevant factors. Some researches have postulated that monetary gains play a crucial role. For example, a growing literature asserts that the role of academics is gradually evolving. Rather than focusing on research with little technological application, academics are more and more eager to connect the worlds of science and technology in an entrepreneurial way, in particular by commercializing the technologies resulting from their research (Clark, 1998; Etzkowitz 2003, Shane 2004).

However, other researchers find that research considerations are the main reason for collaboration between academics and industry (Lee 2000, Meyer-Krahmer and Schmoch 1998). According to this argument, the collaboration is mainly motivated by the desire of academics to advance their own research program. Thus, the interaction with industry could be informed by the desire to obtain funds for graduate students, to access laboratory materials, to acquire knowledge applicable to their own research, to test practical applications and complement research funds (Mansfield 1998).

We then present the following hypothesis:

H8. *The engagement of academics in interactions with industry is motivated by different reasons?*

The literature on university-industry relationships suggests that the motivation factors may be different when academics interact with industry. D'Este and Perkman (2011) identified three main reasons: marketing, learning and access to physical resources and access to finance. In the following, we present a discussion of the hypotheses characterizing each motivation.

University researchers motivated by commercialization assume direct responsibility for the technical development linked to their research activities and, more specifically, undertake the commercial exploitation of their technology or knowledge (Etzkowitz 2003). This logic of industry interaction is based on the literature emphasizing the advent of the entrepreneurial spirit (Clark 1998, Etzkowitz 2003, Shane 2004).

Hence the hypothesis:

H8 a: *The exploitation and commercialization of research results is one of the reasons that guides researchers to collaborate with industry.*

In an even different scenario, academics interact with industry primarily to access the resources they need to continue research. From a conceptual point of view, the logic of access to resources differs from the logics considered above. Like the learning logic, the logic of access to resources is inspired by research-oriented objectives and not by commercial objectives. However, in the case of the logic of access to resources, the research objectives can be achieved without close integration of tasks with industrial researchers. Instead of interactive learning, this interaction logic is therefore characterized by the exchange of resources and reciprocity (Oliver, 1990).

According to this logic, the interaction between the university and the business world allows each party to share resources to pursue its own objectives rather than deriving benefits from the results of joint efforts. The context for this type of arrangement is provided by approaches that conceptualize inter-organizational collaboration induced by reciprocity and by how to access the resources and skills that the partners decide not to produce internally (Powell et al. 1996).

There are two types of resources that academics can access when working with industry. First, funding may be appropriate by providing consulting and contractual research services to industrial partners. Usually, these services are fully paid for by industry, but the academic exploitation of the results could be limited by the absence of scientific novelty or confidentiality issues. The participation of industrial partners can, however, also increase the amount of funding that academics can obtain from government research funding sources. The search for "relevance" has become a major criterion for obtaining funding for academic research (Rip 1994). Academics are encouraged to nominate industrial partners in their funding requests (Behrens and Gray 2001) and to create university-industry research centres (Cohen et al. 1994, Lin and Bozeman 2006).

A second type of resource, even in the absence of funding, is a contribution in kind, such as access to objects, data, equipment and materials. Basic and applied research often relies on the fact that academics can access resources that are generally difficult to obtain in academia. This could for example include specific rare reagents used by biotechnology companies. In engineering, academics often demand "real" data about the manufacturing process or the operation of machinery. Still in other cases, industrial partners could provide access to experimental platforms which would be too costly to build in a university context.

We then propose the following 2 hypotheses:

H8b: *Access to financial resources is a reason for motivation that guides researchers to collaborate*

H8c: *Access to physical resources is a reason for motivation that guides researchers to collaborate*

Sherwood, Butts and Kacar (2004) have argued that universities offer a broad access to a wide variety of research expertise and research infrastructure, while industry offers a broad access to a wide range of expertise in product development/marketing, market knowledge (Sherwood et al. 2004) and employment opportunities for university graduates (Lee and Win 2004; Santoro and Betts 2002). Consequently, universities may be motivated to build relationships with industry to leverage these strengths for mutual benefit hence the assumption:

H8d: *Facilitating employment opportunities for students is a motivational reason that guides researchers to collaborate.*

3.1.2 Impact of Collaboration on Innovation

Orientation towards innovation and innovative activities are necessary for companies to become or remain competitive, especially in a global market where information is widely available and new products and services are continuously introduced (Cakar and Erturk 2010; Madrid-Guijarro et al. 2009, Rosenbusch et al. 2011). In this context, the orientation towards innovation is understood as "... the tendency to integrate and support new ideas, novelty, experimentation and creative processes which can lead to new products, services, technological processes" (Lumpkin and Dess, 1996). Companies with a limited level of innovation are likely to lose customers and market share as their products and services become obsolete. When innovative SMEs succeed in their innovations, a larger share of their sales comes from the new product, which allows them to increase turnover in the medium term (Pett and Wolff 2009) which will be devoted to innovation and therefore they can broaden their core clientele and market share through the creation of market-based value (Bala Subrahmanya Bala Subrahmanya 2015; Love and Roper 2015). Innovation therefore improves financial performance and ultimately contributes to the economic growth of a country (Hausman and Johnston 2014; Karabulut 2015).

In this context, knowledge links, such as that with the university, can facilitate SMEs' access to ideas, by improving knowledge transfer (Lasagni 2012).

For theorists of the Triple Helix Model (MTH), universities play a key role in the innovation system. Knowledge is the focal point of this model and universities play a very important role for knowledge-based societies. In addition, Etzkowitz and Leydesdorff (2000) suggested that universities play a key role in a knowledge-based economy in the generation, development and diffusion of innovation activities. In summary, Etzkowitz (2003) stressed "... that interaction in academia-industry-government is the key to improving the conditions for innovation in a knowledge-based society" Etzkowitz (2003). " the interaction in university-industry-government is the key to improving the conditions for innovation in a knowledge-based society "(Etzkowitz 2003)".

The Triple Helix (MTH) model is based on the importance of relationships between enterprises, government and universities on the transfer of knowledge, a key factor in the development of innovation systems. Etzkowitz (2003) explained that innovation is the basis of MTH, where knowledge plays an important role. An MTH that studies the interaction between the university, industry and government, but primarily focuses on supporting the university in technology-based businesses. Etzkowitz (2003) explained that universities should be involved in training and sharing knowledge processes and that the government should go further than simply playing the traditional role of setting regulations by becoming a public entrepreneur and a venture capitalist.

Several studies have also demonstrated the positive relationship in the interaction between university, industry and innovation (Maietta 2015, Giuliani and Arza 2009; Belderbos et al. Belderbos et al. 2004a, b, Eom and Lee 2009; Santoro 2000; Geisler 2001) They also believe that certain types of interactions are more likely to generate innovation than others.

Therefore, we propose the following hypothesis:

H9: *Collaboration with the university has a positive impact on innovation*

The high relevance of knowledge in the University Business relationship requires control mechanisms due to the ambiguity and contextual specificity of these activities (Becerra et al. 2008; Bouncken and Teichert 2013), therefore, governance of the university business relationship must ensure systematic monitoring and strengthening of relationships in the exchange of inter-organizational knowledge.

Heide (1994) defines governance as "the tools used to establish and structure exchange relationships". They form the institutional and normative framework for inter-organizational relationships and serve both to mitigate the risks of opportunistic behaviour and to coordinate resources (Bosch-Sijtsema and Postma 2009). These mechanisms seek to avoid dysfunctions in collaboration and to coordinate individual contributions (Hoetker and Mellewigt 2009). The literature suggests numerous governance mechanisms such as contracts (Poppo and Zenger 2002), formal directives (Bouncken 2009), reporting mechanisms (Hoetker and Mellewigt 2009), trust (Bosch-Sijtsema and Postma 2009), social ties (Lambe et al. 2001), collaborative activities (Zhou et al. 2015) or relational standards (Tangpong et al. 2010). However, there is a greater consensus on the differentiation between two distinct types of governance: Contractual mechanisms and Relational mechanisms (Bouncken et al. 2016, Cao and Lumineau 2015; Poppo and Zenger 2002; Zhou et al. 2015).

The roots of relational governance mechanisms stem from the theory of social exchange (Lambe et al. 2001). These mechanisms strongly emphasize certain aspects of inherent and moral control, that is to say that they govern exchange activities by pursuing coherent objectives in the context of a cooperative atmosphere (Liu et al. 2009). Compared to contractual governance, relational governance is "closely linked to individuals and it is specific to their relationships" (Hoetker and Mellewigt 2009). It relies heavily on the interaction of people to establish personal ties, strengthen trust and social identification, thus mitigating the risk of opportunism (Bosch-Sijtsema and Postma 2009, Hoetker and Mellewigt 2009). Establishing relational governance involves recurring interpersonal interactions (for example, regular personal meetings, joint problem solving, education and training) (Dyer and Singh 1998 ; Hoetker and Mellewigt 2009).

On the other hand (Smith et al. 2005) define innovation as being the "function of a firm's capacity to create, manage and maintain knowledge" As knowledge is created and stored within individuals (Grant 1996; Grant 1997), human resources and the practices and institutions that influence the value and behaviour of human resources, can play a crucial role in the innovation process (De Winne and Sels 2010). Human resources, in particular their education and training, play an important role in organizations, and there is a growing recognition that human resources are a source of competitive advantage (Laursen and Fos 2003), since resources such as technology are easily imitated by competitors. Innovation in SMEs is particularly affected by the knowledge, skills and capacities of a company's human resources (Laursen and Foss 2003; Mariz-Pérez et al. 2012a, b,), since employees in the company contribute to acquire, develop and exploit new knowledge, which, in turn, contributes to its innovation performance.

We then propose the following hypothesis:

H9a: *Relational mechanisms are positively linked to innovation.*

In addition to the links with the university, the knowledge gained through contractual relationships can also help companies better understand emerging scientific and technological developments and give them easier access to new ideas, creative knowledge and sources of a technical assistance and expertise via collaborative, contractual research and consulting (Lasagni 2012).

According to Bos-Brouwers (2010), SMEs often lack knowledge, so universities can be the missing link in the creation of new innovations, with subject matter experts in the form of university consultants providing a vast amount of knowledge, not previously disclosed. Likewise, Bigliardi et al. (2012) and Klewitz and Hansen (2014) argue that knowledge networks, in particular universities, research centres and consultants, have the knowledge required to provide direct assistance to SMEs. In collaborations between universities and industry in particular, the often tacit nature of advanced scientific and technological knowledge makes interactions particularly relevant (Dornbusch and Neuhäusler 2015). According to Dornbusch and Neuhäusler (2015), R & D collaborations are the privileged means of exchange, allowing regular face to face contacts, reciprocal and bidirectional exchanges of knowledge as well as the circulation of ideas between theory and practice (Perkmann and Walsh 2007).

Yarahmadi and Higgins (2012) report that universities contribute in many ways to business innovation, whether through knowledge leaders or experts who help businesses stay at the forefront of the technological developments to provide businesses with new sources and innovative ideas.

There is an increased pressure on universities to help improve national economic competitiveness (which depends on improvements at the enterprise level) which contributes to their growing involvement in industry, in industrial policy often based on public research institutions stimulating the R & D at the enterprise level (Nieto and Santamaría 2010; Perkmann and Walsh 2007). At the same time, universities are coming under increasing funding pressures, encouraging them to collaborate with the enterprises. Some research indicates that these institutions do more applied research (Nieto and Santamaría 2007). According to Chais et al. (2017), it is necessary that companies and universities understand that they must join their efforts in the collaborative technological research, so that the financial resources invested are not only accepted as articles published in specialized journals, but also transformed into technological innovations accepted by the market. All these investments must come back as new products, services and technologies producing local, regional, national and even international impact, implementing new types of businesses, new markets and generating an economic impact in the country, leading thus to innovation.

We then propose the following hypothesis:

H9b: *Contractual links are positively linked to innovation*

3.2 Conceptual Model

See Fig. 6

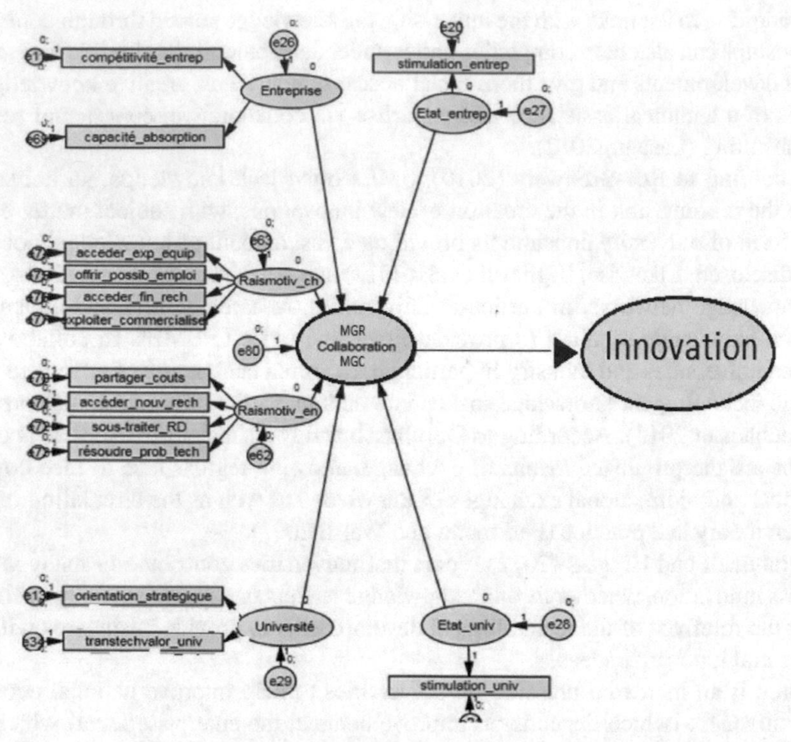

Fig. 6. Conceptual Model: Factors of university enterprise collaboration and its impact on innovation

4 Methodology

4.1 Context of the Research

This study uses the "triple helix" model as a framework for understanding the analysis of the relationship between Tunisian universities and companies belonging to different sectors in the knowledge economy. To meet this objective, several survey methods, including questionnaires and interviews, are used to study the relationship between government, universities (Institutes, schools and faculties belonging to the University of Sfax) and companies located in the region of Sfax.

The choice of the region of Sfax is argued by the fact that:

The governorate of Sfax is located in the Centre-East of the Tunisian Republic, it is bounded by the Mediterranean Sea to the east, by the governorate of Mahdia to the North by the governorate of Kairouan, Sidi Bouzid and Gafsa to the west, and by the governorate of Gabès in the South. It covers 7545 km2 that is 5% of the total area of the country. The economy of the governorate of Sfax, which was previously based essentially on olive oil, fishing and phosphates, has known significant economic changes since the sixties with the proliferation of small and medium-sized manufacturing industries, rapid development of the tertiary sector and diversification of the agricultural sector through the emergence of new investment niches such as cattle breeding, poultry farming and organic crops.

The leading producer in Tunisia, Sfax produces an average of 40% of olive oil and 30% of almond compared to the national level.

Sfax is a large industrial centre (the second after the region of Tunis) which is constantly developing. According to statistics presented by the agency for the promotion of industry and innovation (APII), this industrial centre contains 2300 manufacturing companies employing 59 000 people or 25% of the employed labour force.

Sfax is also a university city with 32985 students enrolled in 20 dynamic establishments forming a network open to its environment (Table 1).

Table 1. Distribution of students from the University of Sfax

University year 2016–2017	2016–2017
Total number of establishments 20	20
Number of Research Structures 120	120
Number of students	32985
Breakdown by gender	
Male	10397
Female	22588
Breakdown by nationality	
Tunisians	32594
Foreigners	391

Source: University of Sfax website

The University of Sfax also has 5 doctoral schools located in different establishments (Table 2).

4.2 Collection of Data

We contacted companies based on a list of companies from different sectors provided to us by:

- The APII (National Agency for Industry and Innovation)
- UPMI (The Union of Small and Medium Enterprises)
- The internship managers of some institutions

Table 2. The five doctoral schools of the University of Sfax

Denominamtion	Establishment
Doctoral school of economic sciences, management and computer science	Faculty of economic sciences and management of Sfax
Doctoral school of letters, arts and humanities	Faculty of letters and human sciences
Doctoral school of science and technology	National school of engineers of Sfax
Doctoral school of legal sciences	Faculty of law of sfax
Doctoral school of basic sciences	Faculty of sciences of sfax

So we initially targeted a sample of around 500 companies. Respondents are generally leaders. In the event of difficulty in accessing managers, we have accepted a senior manager, having information on the company and the company's strategic choices, (technical director or human resources director) to answer our questionnaire.

Finally, we had a return of 250 questionnaires. For the others, the leaders abstained from answering us. Out of these 250 questionnaires and after an in-depth examination, we selected 200 questionnaires which we consider usable. The others were discarded for reasons of the non-completion of the questionnaire.

Regarding the second group that is the researchers, we contacted the researchers relying on personal relationships and on the basis of the contacts provided by the management of some schools and institutes as well as the doctoral schools established in the Faculty of Economics and Management of Sfax, the engineering school of Sfax, the faculty of sciences of Sfax, the faculty of letters and human sciences of Sfax.

Thus, we targeted a sample of around 700 researchers.

The respondents are generally research professors, doctoral students and students engaged in research projects such as Masters students and engineers.

Finally, we received 430 questionnaires. For the others, the researchers abstained from answering us. Out of these 430 questionnaires and after an in-depth examination, we selected 400 questionnaires which we consider exploitable. The others were discarded for reasons of the non-completion of the questionnaire.

The responses are classified on a Likert scale ranging from 1 to 5. This process is preceded by a process of definition and construction of the model variables.

The Table 3 below summarizes the construction of the variables.

Table 3. Representation of variables .

Variables linked to the Enterprise	**Authors**
- The competitiveness of the Enterprise *- Absorption capacity*	- Liao.Sh, Kuo.F et Ding.L (2017). - Borrell-Damian, Morais et Smith (2014) -Garcia-Perez-de-Lema, Madrid-Guijarro et Philippe Martin (2016)
Variables related to industrial motivations	**Authors**
- Access to research *- Solving technical problems* *- The outsourcing of Research and Development - Cost sharing*	- Santoro & Gopalakrishnan (2000) - Belderbos et al. (2004) et Veugelers et al. (2005)
Variables linked to the University	**Authors**
- The strategic orientation of the University - Promotion of technology transfer	- O.Belkhodja et R. Landry (2007) -Chais et al (2017)
Variables related to the researcher's motivations	**Autors**
- Access to financial resources - Access to physical resources - Exploitation and commercialization of research results *- Facilitating employment opportunities for students*	D'Este et Perkman (2011) Lin et Bozeman, (2006)
Variables related to Collaboration Company-University	**Authors**
-Relational and contractual mechanisms	D'Este et Patel, (2007), Bruneel et al. (2010)
Les variables liées à l'Etat	**Authors**
- Stimulation of the State to encourage cooperation	Mohnen et Hoareau, (2003).
Variables related to Innovation	**Authors**
- Different aspects of innovation	Madrid-Guijarro et al. (2009), Estevez(2012).

5 Results

In the course of this fourth and last chapter, we will analyse and discuss the results obtained in order to answer our research question: Can the relationship between business and university with the State assistance facilitate the knowledge enrichment and innovation in Tunisia?

To achieve such an objective, the first section will be devoted to describe our study sample and its characteristics.

In a second section, the results of exploratory factor analyses will be presented, in order to verify the reliability and validity of the measurement scales of the various variables constituting our research model.

The third section will be devoted to confirmatory factor analysis and the testing of hypotheses.

Finally, the fourth and last section will be reserved for the interpretation and discussion of the test results of all the hypotheses of our research, which will allow us to empirically validate or not our conceptual model.

5.1 Statistical Data Processing

The specificities of our research, the problematic, the hypotheses, the nature of the data, led us to use the following data analysis techniques: principal components analysis (PCA); multiple regression. The treatments were used for the analysis of the data, namely SPSS 20 and AMOS 22. All the data issuing from the questionnaire were transmitted in an initial matrix which made it possible to verify and codify the data (Figs. 7 and 8).

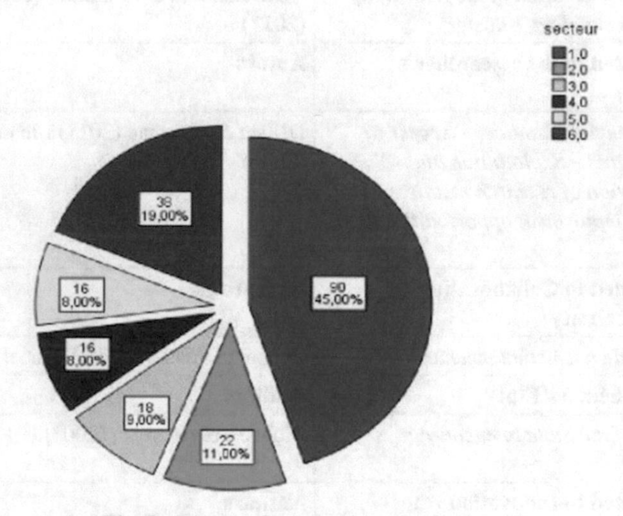

Fig. 7. Distribution of companies by sector

5.2 Descriptive Analysis

5.2.1 Characteristics of Companies

- Sector:

  45% belong to the mechanical, metal, metallurgical and electrical sector, i.e. 90 companies.
  11% belong to the **agricultural and food** industries sector, i.e. 22 companies.
  9% belong to the **chemical** industries sector, i.e. 18 companies.
  8% belong to the **building materials, ceramics and glass** industries, i.e. 16 companies.

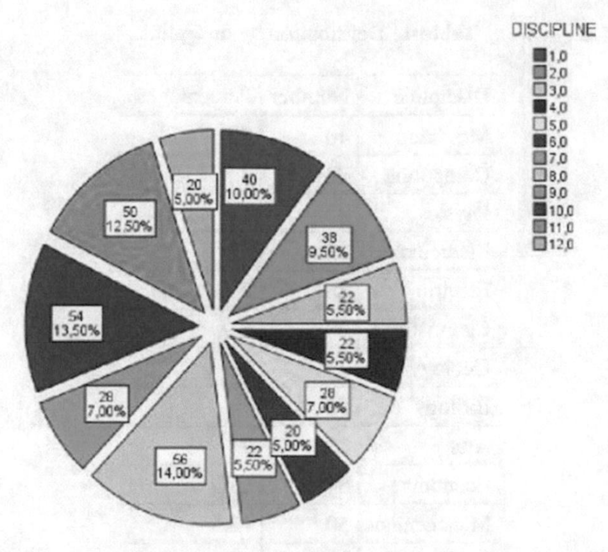

Fig. 8. Distribution by discipline

8% belong to the **Information and Communication Technologies** sector, i.e. 16 companies.
19% belong to other sectors, i.e. 38 companies. (Wood and cork, leather and shoes, textiles and clothing, paper and cardboard).

5.2.2 Characteristics of Researchers

The largest number of respondents belonged to the Biology discipline, i.e. 58 researchers, that is a rate of 14%, then 54 respondents belonged to the Economy discipline, that is 13.5%, then 50 respondents belonged to the Management discipline, i.e. 12.5% and 40 respondents belonged to the Mechanical discipline, i.e. 10% (Table 4).

5.3 Econometric Results

The specificities of our research, the problematic, the hypotheses, the nature of the data, led us to use the following data analysis techniques: principal components analysis (PCA); multiple regression. The treatments were used for the analysis of the data, namely SPSS 20 and AMOS 22. All the data from the questionnaire were transmitted in an initial matrix which made it possible to verify and codify the data.

5.3.1 Measurement and Validation

Anderson and Gerbing (1988) assume the need to assess, first, the dimensionality of the measurement scales and second, the assessment of the internal consistency of each of its dimensions. Thus, they assume that a scale with good internal consistency is not necessarily one-dimensional. Indeed, we proceed, first, to a test of one -dimensionality of the scales then we will study the reliability of the measurements.

Table 4. Distribution by discipline

Discipline	Number	Percentage%
Mecanic	40	10
Computing	38	9.5
Physics	22	5.5
Medicine	22	5.5
Electric	28	7
Chemistry	20	5
Geology	22	5.5
Biology	56	14
Arts	28	7
Economy	54	13.5
Management	50	12.5
Math	20	5

However, before proceeding with the PCA, two tests must be checked: firstly, we need to make sure about the significance of the Bartlett, and Kaiser-Meyer-Olkin (KMO) tests. Second, we perform one-dimensional and reliability tests.

In this research, we used both the KMO and Bartlett's sphericity test to study the relevance of the factor analysis and Cronbach's Alpha indicator to test reliability.

University reliability test

According to several authors, including Evrard et al (1993), the analysis of reliability through Cronbach's alpha indicator is an important step.

The final table in this analysis is the one containing the value of the Cronbach's alpha index. We note here that the value of all the coefficients is greater than 0.7, it then exceeds the minimum required threshold of 0.70 (Nunnaly, 1978) (Tables 5 and 6).

Table 5. Reliability statistics

Cronbach 's alpha	Number of elements
,950	86

Enterprise reliability test

See Tables 7 and 8

5.3.2 Estimation of the Research Model

A multi-varied analysis is used to test the relations between the university and the enterprise and the State and their impact on innovation and on the Entrepreneurial University.

Table 6. Variables/Cronbach's Alpha

Variables	Cronbach's Alpha
University strategic directions	0.795
Promotion of Technology Transfer	0.887
Reasons for motivation	0.915
State stimulation	0.744
Collaboration mechanisms	0.971

Table 7. Reliability statistics

Cronbach's alpha	Number of elements
,920	89

Table 8. Cronbach Variables/Alpha

Variables	Cronbach's alpha
Company competitiveness	0.698
Absorption capacity	0.783
The reasons of motivation of the industrial	0.880
Collaboration mechanisms	0.951
State stimulation	0.752
Innovation	0.705

AMOS is commonly used to analyze multiple regression equations (Nachtigall et al. 2003). The maximum likelihood estimate (MLE) was used as an estimation method and the validity of the construct was assessed. Next, the measurement model with all latent and observed factors included in a full measurement model was examined. Finally, the complete model was analyzed (Table 9).

Table 9. Estimation of the research model

Collaboration	<—	competitiveness_entrep	**0,103**	**0,03**	**3,411**	***
Collaboration	<—	Absorption_cap	0,398	0,036	10,92	***
Collaboration	<—	acceder_exp_equip	0,012	0,036	0,327	**0,744**
Collaboration	<—	offer possib_employment	0,079	0,036	2,165	*0,03*
Collaboration	<—	acceder_fin_rech	0,145	0,036	3,983	***
Collaboration	<—	exploit_market	−0,046	0,036	−1,26	**0,209**
Collaboration	<—	share_costs	−0,111	0,036	3,04	*0,002*
Collaboration	<—	access _ new _ rech	0,125	0,036	3,422	***
Collaboration	<—	subtract RD	0,088	0,036	2,416	*0,016*
Collaboration	<—	solve prob_tech	0,122	0,036	3,338	***
Collaboration	<—	Strategie_univ	0,189	0,036	5,188	***
Collaboration	<—	Transtechvalor _ univ	0,614	0,038	16,3	***
Innovation	<—	GUC	0,029	0,038	0,778	0,436
Innovation	<—	GUR	0,125	0,03	4,144	***

6 Discussions

In our study, we identified several determinants that allow the company as well as the university to establish a collaborative relationship. On the one hand, the competitiveness of the company, the absorption capacity were qualified as being factors which favour the collaboration of the company with the university.On the other hand and from the point of view of the researchers, the strategy of the university, and actions to promote technology transfer encourage cooperation with the company.

In our study, we also tried to confirm that researchers and industrialists have motivational reasons for collaborating and that the latters are finding it increasingly advantageous to collaborate. The most important and significant reasons in relation to manufacturers are the access to research, the resolution of technical problems, the cost sharing and subcontracting of part of the Research and Development activities.

Regarding researchers, the main motivations for collaborating are the access to research funding and access to equipment, and not commercialization.

Finally, as for the impact of collaboration on innovation, we have found that contractual mechanisms such as contractual research and collaborative research are complex, requiring high investments in personnel and equipment. They also require a long period since procedures tend to be more formalized. Thus, a growing formalization and monitoring of the University business relationship could lead to conflicts and mistrust among the parties in their attempt to maintain the autonomy of their organizations in the face of a growing interdependence. Intellectual property issues and how to deal with knowledge are difficult to manage and therefore these mechanisms have not had a positive and significant impact on innovation. On the other hand, relational governance mechanisms such as the transfer of human resources/ mobility, scientific publications and conferences, have a positive impact on innovation.

This result is in line with our hypothesis, which states that relational mechanisms positively affect innovation and confirms the results of (Laursen and Foss 2003, Mariz-Pérez et al. 2012a, b, Sawang and Unsworth 2011) who found that innovation in SMEs is particularly affected by the knowledge, skills and capacities of a company's human resources, given that employees enable a company to acquire, develop and exploit new knowledge, which, in its in turn, contributes to its innovation performance.

As knowledge is created and stored within individuals (Grant, 1996; Grant 1997), human resources as well as practices and institutions influencing the value and behaviour of human resources can play a crucial role in the process of innovation (De Winne and Sels 2010).

We can therefore conclude that the knowledge accessible by transfer of human resources (employment of new graduates, cooperation in the education of graduate students and professional training of employees,) have a positive and significant influence on innovation.

Thus, a key element of this study is that the institutions of higher education play a role in the development and formation of human capital. If universities have long been recognized as providers of education, this role is nevertheless important because the results of this study show that it has a significant positive effect on the results of innovation.

7 Conclusion

Cooperation between universities and industry is not only in the interest of the two institutional partners involved. In an environment where international competition is constantly increasing and the development of technology is very rapid, governments are also interested in good cooperation between the universities and industry, in order to improve the efficiency of innovation, the development of the entrepreneurial university and also to improve the economic development of the country.

In our study, we sought to examine the determinants of University-Enterprise collaboration (EU), to analyse the impact of this collaboration on innovation and to propose a conceptual model in order to explain the relationships between the actors (University -Company- State). Our contribution consists in analysing the cooperation defined by the relational and contractual mechanisms from the angle of the company as well as that of the university while relying on two surveys; one with researchers from the University of Sfax and another with industrialists in the region of Sfax and to study the impact of external collaboration on business innovation, something that has been little discussed in the literature.

To understand and analyse the relationship between universities and companies in the region of Sfax, we opted for the triple helix model, as the main framework of this research thanks to its empirical richness which allows it to be applicable in all countries and because of the weaknesses mentioned with regard to other theoretical models available in the literature, such as the national innovation system (SNI) which supposes that the enterprise is the centre of the innovation process and that the State is the main actor that facilitates this relationship, therefore, it relegates to the second rank the role played by academic institutions in the production of knowledge, and the model of Gibbons et al (1994) which is essentially based on a radical transformation of university-society relations.

Collaboration between universities and industries is therefore essential for the skills development (education and training), generation, acquisition and adoption of knowledge (innovation and technology transfer) and promotion of entrepreneurship (start -up and spin-off). The benefits of university-industry relationships are vast: they can help coordinate R&D programs and avoid overlap, stimulate private investment in R&D and exploit synergies and complementarities of scientific and technological capabilities. University-industry collaboration can also increase the relevance of research carried out in public institutions, promote the commercialization of public R&D results, and increase labour mobility between the public and private sectors. The benefits of university-industry collaboration are also evident in developing countries, and indeed our study has confirmed that collaboration with universities has greatly increased the propensity of companies to innovate.

In general, we can conclude that companies on the one hand are gradually adopting innovation strategies to improve access to and integration of external sources of knowledge, which is generating increased interest in collaboration with universities. And on the other hand, the strategic mission of universities has moved from the tradition of teaching and research to a "third mission" aimed at better meeting the needs of industry and contributing directly to the economic growth and development. The result of this new dynamic is the entrepreneurial university.

References

Articles

Anderson, J.C., Et Gerbing, D.W.: Structural equation modeling in practice: a review and recommended two-step approach. Psychol. Bull. **103**(3), 411–423 (1988)

Audretsch, D.B.: Agglomeration and the location of innovative activity. Oxford Rev. Econ. Policy **14**(2), 18–29 (1998)

Audretsch, D.B., Hülsbeck, M., Lehmann, E.E.: Regional competitiveness, university spillovers, and entrepreneurial activity. Small Bus. Econ. **39**(3), 587–601 (2012)

Bala Subrahmanya, M.H.: Innovation and growth of engineering SMEs in Bangalore: why do only some innovate and only some grow faster? J. Eng. Tech. Manage. **36**, 24–40 (2015)

Behrens, T.R., Gray, D.O.: Unintended consequences of cooperative research: Impact of industry sponsorship on climate for academic freedom and other graduate student outcome. Res. Policy **30**(2), 179–199 (2001)

Belkhodja, O., Landry, R.: The Triple-Helix collaboration: why do researchers collaborate with industry and the government? What are the factors that influence the perceived barriers? Scientometrics **70**(2), 301–332 (2007)

Bellucci, A., Pennacchio, L.: University knowledge and firm innovation: evidence from European countries. J. Technol. Transfer **41**(4), 730–752 (2015)

Becerra, M., Lunnan, R., Huemer, L.: Trustworthiness, risk, and the transfer of tacit and explicit knowledge between alliance partners. J. Manage. Stud. **45**(4), 691–713 (2008)

Bercovitz, J., Feldman, M.: Entrepreneurial universities and technology transfer: a conceptual framework for understanding knowledge-based economic development. J. Technol. Transfer **31**(1), 175–188 (2006)

Bernatchez, J.: De la république de la science à l'économie du savoir: 50 ans de politiques publiques de la recherche universitaire au Québec. Cahiers de la recherche sur l'éducation et les savoirs **11**, 55–72 (2012)

Bishop, K., D'Este, P., Neely, A.: Gaining from interactions with universities: multiple methods for nurturing absorptive capacity. Res. Policy **40**, 30–40 (2011)

Belderbos, R., Carree, M., Diederen, B., Lokshin, B., Veugelers, R.: Heterogeneity in R&D cooperation strategies. Int. J. Ind. Organ. **22**, 1237–1263 (2004a)

Belderbos, R., Carree, M., Lokshin, B.: Cooperative R&D and firm performance. Res. Policy **33**, 1477–1492 (2004b)

Blumenthal, D.: Academic-industrial relationships in the life sciences. N. Engl. J. Med. **349**, 2452–2459 (2003)

Bonarccorsi, A., Piccaluga, A.: A theoretical framework for the evaluation of university-industry relationships. R&D Manage. **24**, 229–247 (1994)

Bouncken, R.B., Kraus, S.: Innovation in knowledge-intensive industries: the double-edged sword of coopetition. J. Bus. Res. **66**(10), 2060–2070 (2013)

Bouncken, R.B., Claußt, T., Fredrich, V.: Product innovation through coopetition in alliances: singular or plural governance? Ind. Mark. Manage. **53**, 77–90 (2016)

Bosch-Sijtsema, P.M., Postma, T.J.: Cooperative innovation projects: capabilities and governance mechanisms. J. Prod. Innov. Manag. **26**(1), 58–70 (2009)

Bos-Brouwers, H.E.J.: Corporate sustainability and innovation in SMEs: evidence of themes and activities in practice. Bus. Strategy Environ. **19**, 417–435 (2010)

Bozeman, B., Rimes, H., Youtie, J.: The evolving state-of-the-art in technology transfer research: revisiting the contingent effectiveness model. Res. Policy **44**(1), 34–49 (2015)

Chais, C., Ganzer, P.P., Olea, P.M.: Technology transfer between universities and companies: two cases of Brazilian universities. Innovation et Management Revue (2017)

Chau, V.S., Gilman, M., Serbanica, C.: Aligning university–industry interactions: the role of boundary spanning in intellectual capital transfer. Technol. Forecast. Soc. Change **123**, 199–209 (2017)

Clark, B.R.: Creating Entrepreneurial Universities: *Organizational Pathways of Transformation*, p. 163p. Pergamon, Oxford (1998)

Clauss, T., Spieth, P.: Treat your suppliers right! Aligning strategic innovation orientation in captive supplier relationships with relational and transactional governance mechanisms. R&D Management (2016)

Cohen, W.M., Levinthal, D.A.: Absorptive capacity: a new perspective on learning and innovation. Adm. Sci. Q. **35**(1), 128–152 (1990)

Cohen, S.B., Florida, R., Coe, W.R.: University-industry partnerships in the US. Pittsburgh, Carnegie-Mellon University (1994)

Corbel, P., Chomienne, H., Serfati, C.: L'appropriation du savoir entre laboratoires publics et entreprises. Revue française de gestion **1**, 149–163 (2011)

Cuijpers, M., Guenter, H., Hussinger, K.: Costs and benefits of inter-departmental innovation collaboration. Res. Policy **40**, 565–575 (2011)

D'Este, P., Patel, P.: University-industry linkages in the UK: What are the factors determining the variety of interactions with industry? Res. Policy **36**(9), 1295–1313 (2007)

D'Este, P., Perkmann, M.: Why do academics engage with industry the entrepreneurial university and individual motivations. J. Technol. Transfer **36**(3), 316–339 (2011)

de Winne, S., Sels, L.: Interrelationships between human capital, HRM and innovation in Belgian start-ups aiming at an innovation strategy. Int. J. Hum. Resour. Manag. **21**, 1863–1883 (2010)

Dornbusch, F., Neuhäusler, P.: Composition of inventor teams and technological progress – the role of collaboration between academia and industry. Res. Policy **44**, 1360–1375 (2015)
Etzkowitz, H., Klofsten, M.: the Innovating region: toward a theory of knowledge-based regional development. R&D Manag. **35**(3), 243–255 (2005)
Etzkowitz, H.: Research groups as 'quasi-firms': the invention of the entrepreneurial university. Res. Policy **32**(1), 109–121 (2003)
Etzkowitz, H.: Triple Helix clusters: boundary permeability at university—industry— government interfaces as a regional innovation strategy. Environ. Plann. C: Govern. Policy **30**(5), 766–779 (2012)
Etzkowitz, H.: Anatomy of the entrepreneurial university. Soc. Sci. Inf. **52**(3), 486–511 (2013)
Etzkowitz, H., Leydesdorff, L.: Introduction to special issue on science policy dimensions of the Triple Helix of university-industry-government relations. Sci. Public Policy **24**(1), 2–5 (1997)
Etzkowitz, L., Leydesdorff, M.: The dynamics of innovation: from national systems and "Model 2" to a triple helix of university industry-government relation. Res. Policy **29**(2), 109–123 (2000)
Garcia-Perez-De-Lema, D., Antonia Madrid-Guijarro, A., Philippe Martin, D.: Influence of university–firm governance on SMEs innovation and performance levels. Technological Forecasting & Social Change (2016)
Gemser, G., Leenders, M.A.A.M.: Managing cross-functional cooperation for new product development success. Long Range Plan. **44**(26e), 41 (2011)
Ghemawat, P.: Finding your strategy in the new landscape. Harvard Bus. Revue. **88**(3), 54–60 (2010)
Gibbons, M., Limoges, C., Nowotny, H., Schwartzman, S., Scott, P., Trow, M.: The New Production of Knowledge: The Dynamics of Science and Research in Contemporary Societies. Sage, London (1994)
Grant, R.M.: Prospering in dynamically competitive environments: organizational capability as knowledge integration. Organ. Sci. **7**, 375–387 (1996)
Grant, R.M.: The knowledge-based view of the firm: implications for management practice. Long Range Plan. **30**, 450–454 (1997)
Geisler, E.: Industry-university technology cooperation: a theory of interorganizational relationships. Technol. Anal. Strateg. Manag. **7**(2), 217–229 (1995)
Guerrero, M., Urbano, D.: The impact of Triple Helix agents on entrepreneurial innovations' performance: na inside look at enterprises located in an emerging economy. Technol. Forecast. Soc. Chang. **119**, 294–309 (2016)
Hoetker, G., Mellewigt, T.: Choice and performance of governance mechanisms: matching alliance governance to asset type. Strateg. Manag. J. **30**(10), 1025–1044 (2009)
Henneberg, S.C., Naude, P., Mouzas, S.: Sense-making and management in business networks—some observations, considerations, and a research agenda. Ind. Mark. Manage. **39**(3), 355–360 (2010)
Heide, J.B.: Interorganizational governance in marketing channels. J. Mark. **58**(1), 71–85 (1994)
Heidenreich, S., Landsperger, J., Spieth, P.: Are innovation networks in need of a conductor? Examining the contribution of network managers in low and high complexity settings. Long Range Plan. **49**, 55–71 (2016)
Hausman, A., Johnston, W.J.: The role of innovation in driving the economy: lessons from the global financial crisis. J. Bus. Res. **67**(1), 2720–2726 (2014)
Jensen, R., Thursby, J., Thursby, M.: Disclosure and licensing of university inventions: "The Best We Can Do with the S&T We Get to Work With. Int. J. Ind. Organ. **21**(9), 1271–1300 (2003)
Lambe, C.J., Wittmann, C.M., Spekman, R.E.: Social exchange theory and re-search on business-to-business relational exchange. J. Bus. Bus. Mark. **8**(3), 1–36 (2001)
Lane, P.J., Salk, J.E., Lyles, M.A.: Absorptive capacity, learning, and performance in international joint ventures. Strateg. Manag. J. **22**(12), 1139–1161 (2001)

Lachmann, J.: Le développement des pôles de compétitivité: quelle implication des universités? Innovations **3**, 105–135 (2010)

Lasagni, A.: How can external relationships enhance innovation in SMEs? New evidence for Europe. J. Small Bus. Manag. **50**, 310–339 (2012)

Lee, J., Win, H.N.: Technology transfer between university research centers and industry in Singapore. Technovation **24**, 433–442 (2004)

Lee, Y.S.: 'Technology transfer' and the research university: a search for the boundaries of university—industry collaboration. Res. Policy **25**, 843–863 (1996)

Lee, S.Y.: The sustainability of university–industry research collaboration: an empirical assessment. J. Technol. Transfer **25**(2), 111–133 (2000)

Liao, S.-H., Kuo, F.-I., Ding, L.-W.: Assessing the influence of supply chain collaboration value innovation, supply chain capability and competitive advantage in Taiwan's networking communication industry. Int. J. Prod. Econ. **191**, 143–153 (2017)

Link, A.N., Paton, D., Siegel, D.S.: An analysis of policy initiatives to promote strategic research partnerships. Res. Policy **31**(8–9), 1459–1466 (2002)

Lin, M.-W., Bozeman, B.: Researchers' industry experience and productivity in university-industry research centers: a "scientific and technical human capital" explanation. J. Technol. Transfer **31**(2), 269–290 (2006)

Liu, Y., Luo, Y., Liu, T.: Governing buyer-supplier relationships through transac-tional and relational mechanisms: Evidence from China. J. Oper. Manag. **27**(4), 294–309 (2009)

Love, J.H., Roper, S.: SME innovation, exporting and growth: a review of existing evidence. Int. Small Bus. J. **33**(1), 28–48 (2015)

Maietta, O.W.: Determinants of university-firm R&D collaboration and its impact on innovation: a perspective from a low-tech industry. Res. Policy **44**(7), 1341–1359 (2015)

Mariz-Pérez, R.M., Teijeiro-Alvarez, M.M., García-Alvarez, M.T.: The importance of human capital in innovation: a system of indicators. In: Soft Computing in Management and Business Economics. Springer, pp. 31–44 (2012)

Mohnen, P., Hoareau, C.: What type of enterprises forges close links with universities and government labs?: Evidence from CIS2. Manag. Decis. Econ. **24**, 133–145 (2003)

Mansfield, E., Hoareau, C.: Academic research and industrial innovation: an update of empirical findings. Res. Policy **26**(7-8), 773–776 (2003)

Mallowan, M., Liquete, V., Verlaet, L.: De la gestion des connaissances à l'économie des connaissances. Commun. Manag. **12**(1), 5–12 (2015)

Muscio, A.: What drives the university use of technology transfer offices? Evidence from Italy. J. Tech. Transfer. **35**, 181–202 (2010)

Muriithi, P., Horner, D., Pemberton, L., Wao, H.: Factors influencing research collaborations in Kenyan universities. Res. Policy **47**(1), 88–97 (2018)

Narula, R.: R&D collaboration by SMEs: new opportunities and limitations in the face of globalisation. Technovation **24**, 153–161 (2004)

Nieto, M.J., Santamaria, L.: The importance of diverse collaborative networks for the novelty of product innovation. Technovation **27**, 367–377 (2004)

Perkmann, M., Walsh, K.: University-industry relationships and open innovation: towards a research agenda. Int. J. Manag. Rev. **9**(4), 259–280 (2007)

Perkmann, M., Walsh, K.: The two faces of collaboration: impacts of universityindustry relations on public research. Ind. Corp. Change **18**(6), 1033–1065 (2009)

Pett, T.L., Wolff, J.A.: SME opportunity for growth or profit: what is the role of product and process improvement?. J. Int. Entrepreneurship **1**(1), 5–21 (2009)

Richardson, G.B.: The organization of industry. Econ. J. **82**(327), 883–896 (1972)

RIP, A.: The republic of science in the 1990s. High. Educ. **V28**(1), 3–23 (1994)

Sawang, S., Unsworth, K.: "Why adopt now? Multiple case studies and surveystudies comparing small, medium and large firms. Technovation **31**(10–11), 554–559 (2011)

Shane, S.A.: Academic Entrepreneurship: University Spinoffs and Wealth Creation. Edward Elgar, Cheltenham (2004)
Sherwood, A.L., Butts, S.B., Kacar, S.L.: Partnering for knowledge: a learning framework for university-industry collaboration. In: Midwest Academy of Management, 2004 Annual Meeting, pp. 1–17 (2004)
Siegel, D.S., Waldman, D.A., Atwater, L.E., Link, A.N.: Commercial knowledge transfers from universities to firms: improving the effectiveness of university–industry collaboration. J. High Technol. Manage. Res. **14**(1), 111–133 (2003a)
Siegel, D.S., Waldman, D., Link, A.S.: Assessing the impact of organizational practices on the relative productivity of university technology transfer offices: an exploratory study. Res. Policy **32**, 27–48 (2003b)
Santoro, M.D., Betts, S.C.: Making industry-university partnerships work. Res. Technol. Manag. **45**, 42–46 (2002)
Tangpong, C., Hung, K.-T., Ro, Y.K.: The interaction effect of relational norms and agent cooperativeness on opportunism in buyer-supplier relationships. J. Oper. Manag. **28**(5), 398–414 (2010)
Veugelers, R., Cassiman, B.: R&D cooperation between firms and universities: some empirical evidence from belgian manufacturing. Int. J. Ind. Organ. **23**, 355–379 (2005)
Yarahmadi, M., Higgins, P.G.: Cooperation as a driver of environmental innovation in Australian businesses. In: Proceedings of the XXIII ISPIM Conference–Action for Innovation: Innovating from Experience (2012)
Zhou, K.Z., Li, C.B.: How strategic orientations influence the building of dynamic capability in emerging economies. J. Business. Res. **63**, 224–231 (2015)
Zhou, Y., Zhang, X., Zhuang, G., Zhou, N.: Relational norms and collaborative activities: Roles in reducing opportunism in marketing channels. Ind. Mark. Manage. **46**, 147–159 (2015)

Books

Evrard, Y., Pras, B., Et Roux, E.: Market, Etudes et recherches en marketing, Paris, Nathan, 1ére édition (1993)
Lemaitre, D.: Formation des ingénieurs à l'innovation. Collection Innovation, Entrepreneuriat, et Gestion. Edition ISTE (2018)
Forest, J.: Petite histoire des modèles d'innovation. Principes d'économie de l'innovation (2014). 514- P
De Frascati, M.: Lignes directrices pour le recueil et la communication des données sur la recherche et le développement expérimental. Dans la série: Mesurer les activités scientifiques, technologiques et d'innovation (2015)
Schumpeter, J.A.: Capitalism, Socialism and Democracy, New York, Harper & brothers, éd. Française Capitalisme, Socialisme et Démocratie, Paris, petite Bibliothèque. Payot, 1951, Rééd. 1974 (1942)

Reports

Conseil De La Science Et De La Technologie (CST): Pour une politique québécoise de l'innovation. Rapport de conjoncture 1998. Québec: Gouvernement du Québec (1997)
FQPPU: La commercialisation de la recherche et de l'expertise universitaire dans les universités québécoises, Comité ad hoc de la Fédération québécoise des professeures et professeurs d'université sur la commercialisation de la recherche, Montréal: les cahiers de la FQPPU (2000)

FQPPU: La propriété intellectuelle en milieu universitaire au Québec, Comité ad hoc sur la propriété intellectuelle de la Fédération québécoise des professeures et professeurs d'université, Montréal : les cahiers de la FQPPU (2002)

OCDE: La mesure des activités scientifiques et technologique. Principes directeurs pour le recueil l'interprétation des données sur l'Innovation. Manuel d'Oslo. 3ème Edition, OCDE (2005)

Author Index

K. Boussafi et al. (Eds.): MSENTS 2019, LNNS 162, p. 123, 2021.
https://doi.org/10.1007/978-3-030-60933-7

Printed in the United States
By Bookmasters

Printed in the United States
By Bookmasters